建筑立场系列丛书 No.21

内在丰富性建筑
Introvert Potential

中文版
（韩国C3 第337期）

韩国C3出版公社 | 编

王思锐 高文 赵敏 时跃 孙铭泽 原满 王慧 | 译

大连理工大学出版社

资讯

- 004 阿尔梅勒2022年花展_MVRDV
- 008 地拉那市中心整体规划_Grimshaw Architects
- 009 瓜德罗普岛Beauperthuy医院_SCAU
- 010 Riviera-Chablais医院_Groupe-6
- 012 集合/联合，2012罗马国立当代美术馆青年建筑师项目_Urban Movement Design

内在丰富性建筑

- 016 *内在丰富性的探索*_Diego Terna
- 022 Brufe社区中心_Imago
- 032 S住宅_Suga Atelier
- 046 日光住宅_Takeshi Hosaka Architects
- 058 Mecenat美术馆_Naf Architect & Design
- 068 Moliner住宅_Alberto Campo Baeza

建筑立场系列丛书 No.21

城市设计
城市遗产

- 080 *城市遗产的填充*_Michele Stramezzi + Maria Pedal
- 086 贝尔法斯特都市艺术中心_Hackett Hall McKnight
- 098 隆德大教堂广场建筑_Carmen Izquierdo
- 104 圣安东慈善学校修复工程_Gonzalo Moure Architect

承孝相

- 120 *融入自然的小山村*_BongHee Jeon
- 126 土、水、花、风，360°乡村俱乐部
- 140 朝韩非军事区和平生活谷
- 152 济州岛艺术别墅社区中心
- 162 "MoHeon"与"SaYaWon"

- 178 建筑师索引

News
004 Almere Floriade 2022 _ MVRDV
008 Central Tirana Masterplan _ Grimshaw Architects
009 Guadeloupe Beauperthuy Hospital _ SCAU
010 Riviera-Chablais Hospital _ Groupe-6
012 Unire/Unite, the 2012 YAP MAXXI _ Urban Movement Design

Introvert Potential

016 *Introversion, or about Observation (within)* _ *Diego Terna*
022 Brufe Social Center _ Imago
032 House S _ Suga Atelier
046 Daylight House _ Takeshi Hosaka Architects
058 Mecenat Art Museum _ Naf Architect & Design
068 Moliner House _ Alberto Campo Baeza

Urban How
City Inherit

080 *Heritage Infills* _ *Michele Stramezzi + Maria Pedal*
086 MAC, Belfast _ Hackett Hall McKnight
098 Lund Cathedral Forum _ Carmen Izquierdo
104 San Antón Charity School Restoration _ Gonzalo Moure Architect

Seung, H-Sang

120 *A Small Village becomes one with Natures* _ *BongHee Jeon*
126 Earth, Water, Flower, Wind, 360° Country Club
140 Korea DMZ Peace and Life Valley
152 Jeju Art Villas Community Center
162 MoHeon and SaYaWon

178 Index

可持续发展城市 SUSTAINABLE CITY

阿尔梅勒2022年花展 _MVRDV

荷兰著名的园艺博览会每十年才举办一次,经过激烈竞争,阿尔梅勒市成为荷兰四座候选城市之一。

MVRDV建筑事务所对阿尔梅勒市的规划,不是从博览会举办地这一时之用的角度出发,而是将其视为现有城市中心的绿色延伸,打造成可持续发展的绿色理想城市。城市中心对面的滨海地区将发展为充满活力的全新城市社区和一座巨大的植物博物馆,博览会结束后也将一直保留。

设计者大胆构想,希望创造出的展览馆无论是从字面意义上的绿色还是从可持续发展角度来看,都能达到目前绿色标准的300%:该地区每个项目都将与植物结合,创造出惊喜,不断创新并达到生态平衡。

阿姆斯特丹的都市区正面临着住房量激增的趋势。阿尔梅勒市拥有六万所新住房,将成为新开发项目的最大受益者。

这一规划预见了高密度展览和绿色城市中心的扩展,同时也极具灵活性:邀请花展承办方参与进一步规划。

Winy Maas在讨论规划时说:"我们梦想打造的绿色城市,不仅是字面意义上的绿色,也包括生态绿色。一座能够生产所需的食物、能源、净化所需的水、废品循环并保持生态多样性的城市。"

阿尔梅勒花展将在45万平方米的半岛上铺设成植物的海洋。每个街区专属于不同的植物,形成按首字母顺序排列的植物博物馆。同时,街区也专属于不同的项目,从凉亭到住宅、办公室甚至大学校园,都将按序规划,并安排成堆叠式植物园,形成一个垂直的生态系统,每类空间拥有不同的气候条件,满足不同植物的生长需求。游客可选择住在茉莉花宾馆,在百合花装饰的游泳池中游泳,在玫瑰园中进餐。这座城市将住宅安排在果园中,在办公室内种植植物,另外还建有竹林园。博览会和新的城市中心将成为提供食物和能源的地方,即一片绿色的城区,它将淋漓尽致地向人们展示植物是如何丰富日常生活各个方面的。

Almere Floriade 2022

Almere is one of the four Dutch cities left in the race for the prestigious horticultural Expo which takes place once every ten years in the Netherlands.
The MVRDV plan for Almere is not a temporary Expo site but a lasting green ideal city as a green extension of the existing city center. The waterfront site opposite the city center will be developed as vibrant new urban neighbourhood and gi-

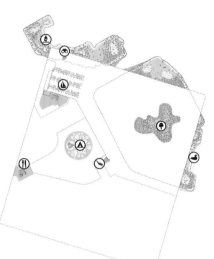

原始森林	primitive woods
水上运动	watersports
观察塔"乌托邦"	watch tower "Utopia"
野营区	camping
餐厅	restaurant
潜水中心	diving center
海滩	beaches
生态芦苇河岸	ecological reed banks

运动场 sports　　观察塔 watch tower

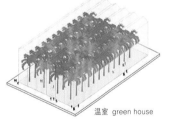
温室 green house

展馆 pavilion

办公楼 office

住宅 homes

酒店 hotel

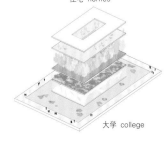
大学 college

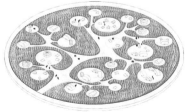
野营区 camping

项目名称：Almere Floriade 2022
地点：Almere, the Netherlands
建筑师：MVRDV
项目团队：Winy Maas, Jacob van Rijs, Nathalie de Vries, Jeroen Zuidgeest, Klaas Hofman, Mick van Germert, Elien Deceuninck, Monika Kowaluk
甲方：City of Almere
设计时间：2012

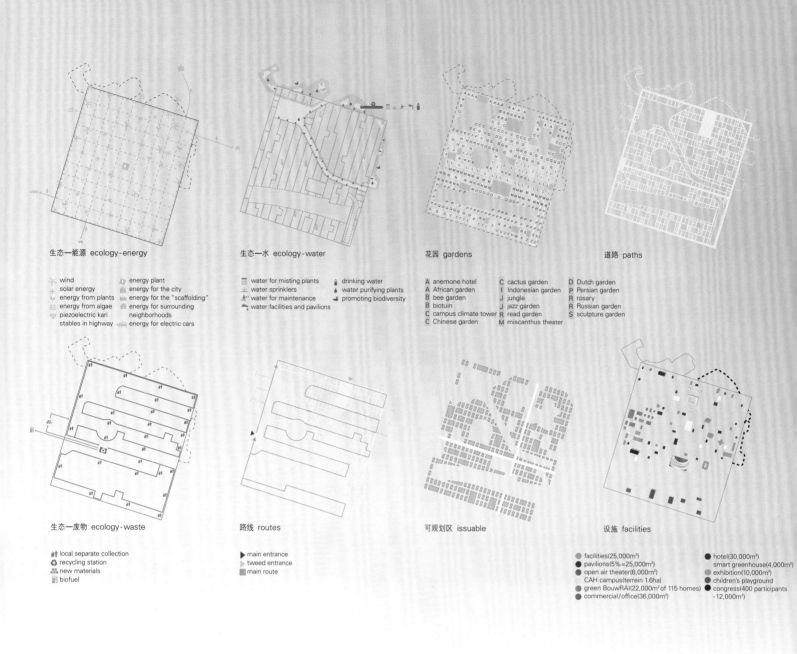

生态—能源 ecology-energy | 生态—水 ecology-water | 花园 gardens | 道路 paths

- wind
- solar energy
- energy from plants
- energy from algae
- piezoelectric kari
- stables in highway
- energy plant
- energy for the city
- energy for the "scaffolding"
- energy for surrounding neighborhoods
- energy for electric cars

- water for misting plants
- water sprinklers
- water for maintenance
- water facilities and pavilions
- drinking water
- water purifying plants
- promoting biodiversity

A anemone hotel
A African garden
B bee garden
B biotuin
C campus climate tower
C Chinese garden
C cactus garden
I Indonesian garden
J jungle
J jazz garden
R read garden
M miscanthus theater
D Dutch garden
P Persian garden
R rosary
R Russian garden
S sculpture garden

生态—废物 ecology-waste | 路线 routes | 可规划区 issuable | 设施 facilities

- local separate collection
- recycling station
- new materials
- biofuel

- main entrance
- tweed entrance
- main route

- facilities(25,000m²)
- pavilions(5%=25,000m²)
- open air theater(6,000m²)
- CAH campus(terrein 1.6ha)
- green BouwRAI(22,000m² of 115 homes)
- commercial/office(36,000m²)
- hotel(30,000m²)
- smart greenhouse(4,000m²)
- exhibition(10,000m²)
- children's playground
- congress(400 participants - 12,000m²)

ant plant library which will remain after the Expo. The ambition is to create a 300% greener exhibition than currently standard, both literally green and sustainable: each program on the site will be combined with plants which will create programmatic surprises, innovation and ecology.

Amsterdam's metropolitan area stands at the verge of a large housing growth. With 60,000 new homes the city of Almere will realize the largest share of this new development.

The plan foresees a dense exemplary and green city center extension which at the same time is very flexible: an invitation to the Floriade organizer to develop the plan further.

Winy Maas discusses the plan, "We dream of making green cities, the city that is literally green as well as ecological. A city that produces food and energy, cleans its own water, recycles waste and holds a great biodiversity..."

Almere Floriade will be developed as a tapestry of gardens on a 45ha square-shaped peninsula. Each block will be devoted to different plants, a plant library with perhaps an alphabetical order. The blocks are also devoted to program, from pavilions to homes, offices and even a university which will be organized as a stacked botanical garden, a vertical eco-system in which each class room will have a different climate to grow certain plants. Visitors will be able to stay in a jasmine hotel, swim in a lily pond and dine in a rosary. The city will offer homes in orchards, offices with planted interiors and bamboo parks. The Expo and new city center will be a place that produces food and energy, a green urban district which shows in great detail how plants enrich every aspect of daily life.

地块概念图 plot concept

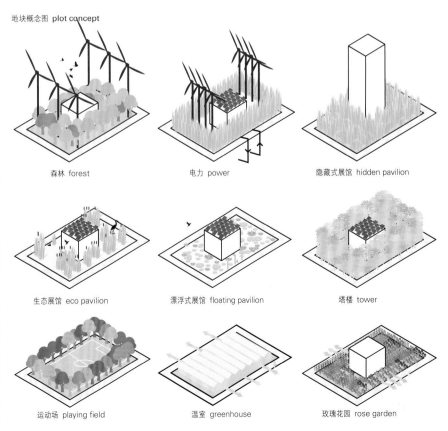

森林 forest　　电力 power　　隐藏式展馆 hidden pavilion

生态展馆 eco pavilion　　漂浮式展馆 floating pavilion　　塔楼 tower

运动场 playing field　　温室 greenhouse　　玫瑰花园 rose garden

地拉那市中心整体规划 _Grimshaw Architects

Grimshaw建筑事务所极其兴奋地宣布，该事务所被选定负责阿尔巴尼亚首都地拉那大规模扩建的总体规划。

地拉那市长要求，设计规划需将20世纪30年代的林荫道延长3km，并新建一座7km长的河滨公园，在此过程中，恢复地拉那河和整座城市往日的活力。项目覆盖了14km²的区域，占整座城市面积的四分之一还多。在欧洲大陆国家的首都进行如此大规模的城市总体规划，绝对是非常难得的机遇。

有若干团队进入第一轮评选，主要有West 8、GMP、Cino Zuccchi、KCAP和Albert Spear & Partners等建筑事务所。经过地拉那市政府和一个国际评审团的评选，Grimshaw和DAR Group两家事务所的设计进入了第二轮角逐。最终阶段的评选包括民意调查和在国家电视台进行展示，在进行民意调查时，地拉那市民可看到两种方案的成比例模型，并留下意见。

Grimshaw建筑事务所的方案将不同的公共空间有序地贯穿在整条林荫道两侧，从几何学上呼应了现有的土地所有权格局。地拉那是一座倡导户外文化的地中海城市。每个区域空间或"活动场所"都展示着不同的功能和特点，构成了全新的"广场交响曲"。

河滨公园也基本按相同的理念规划，但体现了地拉那河的特征。地那拉河是一条"年轻"的山区河流，四季河水流量分明。Grimshaw的设计师们探索了该城市中心文化和自然的独特交融，这成为其城市设计的关键元素。

虽然整座城市的总体规划区域需要很多年来完成，但地拉那市长希望尽快完成林荫大道及其"城市客厅"的建造，使之成为整个项目的脊梁，未来所有的新设施都将围绕这一主干建设。

Central Tirana Masterplan

Grimshaw is thrilled to announce it has been selected to masterplan a large expansion to Albania's capital city, Tirana.

The mayor of Tirana called for designs to extend the 1930s boulevard a further 3km and establish a new 7km riverside park, rejuvenating the river Tirana and the city in the process. The project area covers 14 km², over a quarter the size of the city area, and represents a unique opportunity for a masterplan on this scale in a capital within continental Europe.

Several teams were selected for the first stage, led by West 8, GMP, Cino Zucchi Architects, KCAP and Albert Spear & Partners. After deliberations by the Tirana municipality and an international jury, designs by Grimshaw and the DAR Group were selected for a second stage. The final phase involved both public consultation, where the citizens of Tirana were able to view scale models of both schemes and leave feedback, and presentations broadcast on national television. Grimshaw proposed a robust sequence of public spaces threaded along the boulevard that responded geometrically to existing patterns of land ownership. Tirana is a Mediterranean city with an outdoor culture. Each space or "living room" was presented with a different use and character; a new "symphony of squares".

The river park is structured around the same concept turned through 90 degrees but informed by the character of the river Tirana: a "young" mountain river with a wide seasonal variation of water flow. Grimshaw explored the unique juxtaposition of culture and nature in the centre of the city as key elements of the urban design.

Although the complete masterplan area will take many years to develop, the mayor of Tirana wishes to build the boulevard and its "living rooms" as quickly as possible to act as the backbone around which future development will emerge.

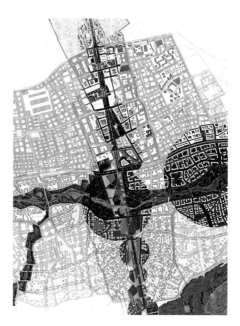

项目名称：Central Tirana Masterplan 地点：Tirana, Albania
建筑师：Grimshaw Architects 合伙人：Neven Sidor
副总监：Ben Heath 参与设计建筑师：Paolo Vimercati
用地面积：14km²

与自然的融合 NATURIZING

瓜德罗普岛Beauperthuy医院 _SCAU

该医院面朝加勒比海，沿水平花园在连续的阶地中布局，便于患者进入。

受到当地风俗和热带建筑构造的启发，该建筑采用生物气候学的设计理念，使其受益于信风带来的自然通风。

该建筑采用了独立亭阁的布局形式，将现代建筑风格与当地的建筑规模和传统融合起来。

Guadeloupe Beauperthuy Hospital
Facing the Caribbean Sea, the hospital is laid out in successive terraces along horizontal gardens that are easily accessible for patients.
A bioclimatic design inspired by local practices and tropical architecture allows it to benefit from the natural ventilation of the trade winds.
The plan in the form of independent pavilions marries a contemporary architectural style with local scale and traditions.

项目名称：Guadeloupe Beauperthuy Hospital
地点：Pointe-Noire, Guadeloupe
建筑师：SCAU
甲方：Centre Hospitalier Louis Daniel Beauperthuy
用途：275 bed hospital
净表面积：19,000m²
造价：EUR 460,000,000
预计竣工时间：2013

A-A' 剖面图 section A-A'

西北立面 north-west elevation

Riviera – Chablais医院 _ Groupe-6

沃州和瓦莱州决定建立一所共用的医院,这在瑞士历史上是前所未有的。医院选址于勒纳镇,因为其靠近高速公路等基础设施,并且位于两州边界这一有象征意义的地理位置上。

医院采用了三层楼高的水平层形式,这一高度使其避免受到相邻村庄较高建筑的视觉冲击。

为了与周围景色相协调,该设计融合了当地的自然地形特征。

住院区集中在建筑顶部的高层上。监护病房朝向周围乡村,能够远眺花园,并避免受到附近高速公路的噪音影响。其他医疗活动都安排在第一和第二层,封闭在反光的防护玻璃墙内。

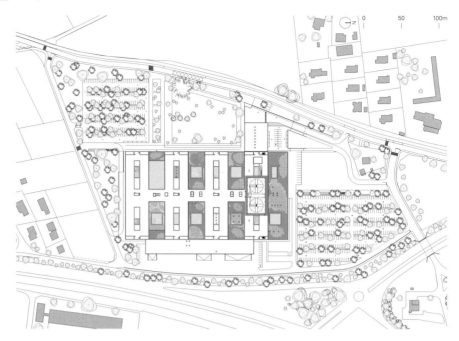

Riviera - Chablais Hospital

The cantons of Vaud and Valais have decided to build a shared hospital, an unprecedented event in Switzerland's history. The site, located in the village of Rennaz, was chosen for its proximity to the motorway infrastructures and its symbolic geographical location on the border of the two cantons.

The hospital takes the form of an horizontal layer rising up over three levels, a height determined to avoid any visual clash with the taller buildings in the neighbouring village.

In harmony with the landscape, it integrates the natural topography of the site.

The in-patient area is concentrated on the upper floor at the top of the building. The patient care units overlook gardens protected from the noise nuisances of the nearby motorways and give onto the surrounding countryside. The rest of the medical activities are located on the first two levels, enclosed within a protective and reflective glass skin.

项目名称:Riviera-Chablais Hospital
地点:Rennaz, Switzerland
建筑师:Groupe-6
项目团队:Geninasca Delefortrie, Quartal, Biol Conseils, TP SA, HVAC FM
结构工程师:Daniel Willi SA
电气工程师:Betelec SA
甲方:Conseil d'Etablissement Hôpital Riviera-Chablais, Vaud-Valais
用途:Acute care and emergency treatment establishment
净表面积:60,000m²
预计竣工时间:2016

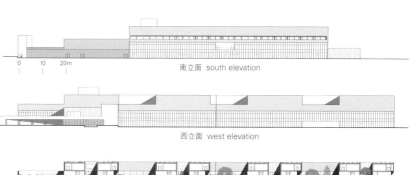

南立面 south elevation

西立面 west elevation

A-A' 剖面图 section A-A'

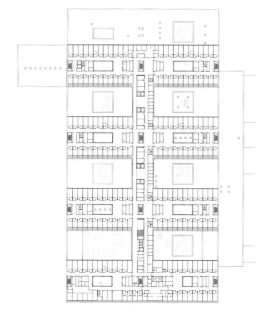

三层 third floor

一层 first floor

集合/联合，2012罗马国立当代美术馆青年建筑师项目 _Urban Movement Design

现代艺术博物馆（MoMA）、纽约MoMA PS1艺术空间与罗马国立当代美术馆（MAXXI）联合开展了罗马国立当代美术馆青年建筑师项目（YAP MAXXI），这是首个在意大利开展的青年建筑师项目（YAP）。

YAP向青年设计师提供机会，让他们在纽约MoMA PS1大庭院和MAXXI广场上设计和创造出用于举办夏季活动的空间，YAP正是因此而闻名。MoMA和MAXXI将Urban Movement Design建筑事务所选为2012年YAP MAXXI的赢家。

集合/联合_Urban Movement Design建筑事务所

我们的设计希望解决通过可进入性、移动性和可持续性这三面透镜发现的问题。我们对待设计上的问题都像对待需要治疗的病人身体，治疗的依据并非一成不变的规则，而是具体问题具体分析。集合/联合这一项目尝试着平衡我们共同的身体，并给我们的生活带来愉悦。

我们建筑的所在地——无极限系统——位于MAXXI的顶部。这是一个阶梯式区域，楼梯介于斜坡和台阶之间，这样的设计保证了坐轮椅和其他有肢体障碍的人都能欣赏到这里的美景。

集合（Unire）的意思是我们都是相互联系的，如同一个肌体内的独立细胞，都是为了整个生物体的融合和健康。我们用缓斜坡代替了台阶，从而使得所有人都可进入这一区域，并融入其中。我们也开辟了一条小路，采用可循环橡胶铺设的软地面，便于轮椅、步行者、婴儿手推车或拄拐杖的人从上面经过。

下了主通道便是无极限系统的快速中转入口，连续的单一表面，双倍无限空间，使得所有用户都处在同一表面上，就如同地球母亲将所有人都拥在她的怀里一样。

无极限系统是为健康运动和休闲活动而设计的，给人们提供不同的场地和项目来强身健体、排出有害物质，并平衡身体系统，从而增强体质，增加身心联系。

为了在这个崭新的全球化世界中保持可持续性，我们必须从现在起，解放思想，敞开心扉，以全球化的角度来审视自身，但也要感受到的是，地球仿佛就是一个村庄，每个人都和其他人以及自然相互联系。我们在主广场的地面上写着"集合/联合"，借此强调这种联系。从水平角度是看不到这些字的，只有从博物馆上方二楼展厅才能看到并完整地了解它们的意图。这意味着我们必须超越自身的渺小，进入统一和相互联系的层面。

Unire / Unite, the 2012 YAP[Young Architects Program] MAXXI

The Museum of Modern Art (MoMA) with MoMA PS1 in New York and the National Museum of Contemporary Art of Rome(MAXXI) have joined forces to launch YAP MAXXI, the first Italian edition of the Young Architects Program(YAP).

YAP is well known for offering young designers the opportunities to design and create a space for live summer events in the great courtyard of MoMA PS1 in New York and in the square of the MAXXI.

MoMA and MAXXI have selected Urban Movement Design as the winner of 2012 YAP MAXXI.

Unire/Unite_Urban Movement Design

Our design solutions look to answer the questions found through the three part lens of Accessibility, Mobility and Sustainability. We treat each design problem as though it is a sick body in need of healing. Not according to any set rule but only according to the particular trauma or need that the body displays at any given time.

The project Unire/Unite attempts to bring

balance to our mutual common body and to breathe joy into our lives.

The site of our built installation, the Infinity System, is located at the top of the MAXXI site, a stepped area with stairs sitting between the ramp and the steps, whereby leaving this beautiful area uninhabited by those in wheelchairs and others with physical impairments.

Unire means we are all connected, as though individual cells in one body, all with the need for the common goal of integration and well-being for the health of the entire organism. We have created a low level slope in place of steps in order to make the area fully accessible and inclusive to all. We have then carved out a pathway, with a soft ground of recycled rubber, easy on the body and easily maneuverable for a wheelchair, walker, stroller or walking stick.

Off the main pathway are easy transfer entrances onto the Infinity, a continuous single surface double infinity which holds all users on one surface, as mother earth holds all of humanity in its arms.

The Infinity System, designed for healthy movement and relaxation, offers varying positions and exercises for the body to take form while activating, strengthening, cleansing and balancing the body's systems for increased health and mind-body connection.

In order to be sustainable in this new global world, we must now open our minds and hearts to feel and view ourselves as a global world, but feel as though we are a small village, each of us interconnected with the others and with nature. We tried to emphasize this by writing "Unire/Unite" on the ground of the main piazza. It is unreadable from eye level, but only from above, from within the second floor gallery of the museum can recognize the words and its intention be read in full. This represents that we must rise above our smallness, and enter into the place of oneness and connection.

项目名称：Unire/Unite
地点：Rome, Italy
设计：Sarah Gluck, Robyne Kassen, Simone Zbudil Bonatti
合作者：Anna Maria Zandara, Andrea Ribechini, Daniele Ludovisi, Marta Veltri, Azzurra Galanti, Vincenzo Gagliardi, Daniele Lampis, Emanuela Magnani, Eduardo Marques, Michael Caton, Ilana Judah, Kerim Eken, Ezra Ardolino, Annah Kassen
摄影师：courtesy of MAXXI

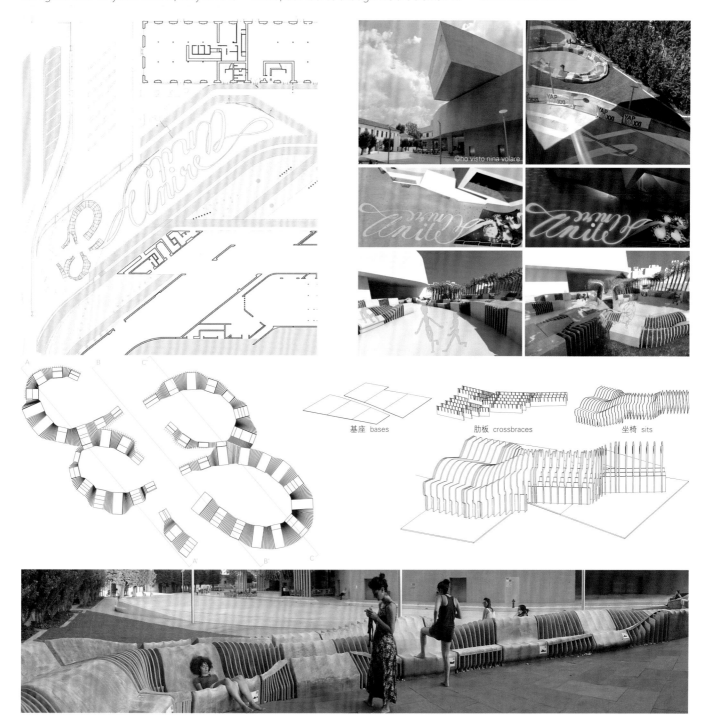

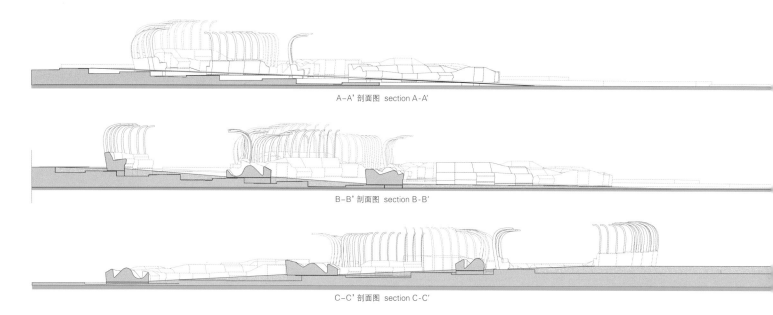

A-A' 剖面图 section A-A'

B-B' 剖面图 section B-B'

C-C' 剖面图 section C-C'

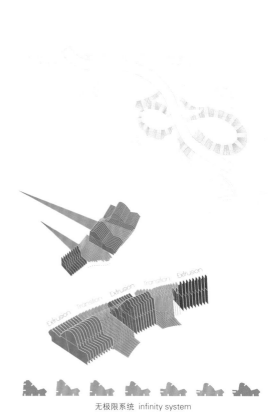

无极限系统 infinity system

屋顶细部 roof detail

内在丰富性建筑
Introvert Potential

贝丝•吉本（全球著名乐队Portishead的主唱）在演唱到歌曲Threads的最后几句时，发出了歇斯底里的呐喊，维罗纳城中斯卡利杰罗城堡的墙壁在这个看似娇弱的女人迸发出的爆发力下仿佛要迸裂一般。从外面看，墙壁非常坚固而无法穿透，在内部看起来却舒适宜人，以歌曲的声音穿透力与听众产生共鸣，成为城堡和城市之间通道的薄膜结构。

火山落灰破坏了原罗马庞贝古城的城市结构，但也将其完好地保存了下来，这让我们直观地看到以手工艺品形态呈现出来的房屋，同时也展现出房屋如何留给人们一个相对隐蔽的私人生活空间：城市中房屋的外部空间是无声的，而内部绿荫环绕，蕴含着水景和多视角元素，尽管它们仍保持相对私密的状态，然而所有这些还是让人们能够在这里欣赏到远方的景色。

不知何故，房屋中的生活却因为那些相同的限制而宣告结束：埃托雷•斯科拉（意大利导演）于1987年导演的影片《家庭》，表现了几代人长时间以来在一条几米长的走廊里不断地进行各种活动，最后这些活动都不复存在的画面。

同样的故事也发生在波尔图的音乐之家里，该建筑如同一个巨大的天外来客不知从何处降临到了这座葡萄牙城市中，屋顶上的一个细微开口展现了建筑柔和婉约的特性。屋顶有一个天井，上面覆盖着黑白两色的瓦片，它将游客带到了小巧的葡萄牙街道和房屋中，然而却没有在环境方面呈现出建筑物的内向性。

我们将从这方面来分析建筑物，讲述这些看似由坚不可摧的墙壁构成的丰富复杂的世界，让它们摆脱萨特在其作品《墙》中塑造的那些噩梦般的场景，而与外部世界产生共鸣，譬如采纳吉本的歌词所传递出的信息，效仿库哈斯使用过的陶瓷材料，来讲述发生在墙内的生活故事，就像世代流传的史诗一样。

Beth Gibbons explodes into an extreme and furious scream during the final bars of *Threads*: the walls of the Scaligero Castle, in Verona, seem to explode with the force bursting from this seemingly fragile woman. The walls, solid and impenetrable from the outside, welcoming on the inside, vibrate at the song's sonic power, becoming a membrane of passage between the castle and the city.

The urban fabric of the original Roman city of Pompeii, destroyed yet preserved by volcanic ash-fall, gives us a vision of the house as an artifact, revealing how the house allowed a quite well-concealed private life: so much so that the outside was mute to the city and the interior was enriched with greenery, water features, and multiple views, all of which gave the rooms a suggestion of overlooking, though they remained quite intimate.

Somehow, life in the houses ended within those same limits: the movie *The Family*, by Ettore Scola, 1987, suggests the exhaustion of all the activities of several generations which had long been conducted along a few meters of a long corridor.

Something similar happens in Porto's Casa da Mùsica, a giant alien object that seems to have dropped from nowhere upon the Portuguese city, revealing its gentle character via a tiny cut in the roof, on which a patio, tiled in black and white, takes visitors to the minute scale of the Portuguese streets and houses, losing the appearance of introversion with respect to the environment. The buildings to be analyzed herein tell of rich and complex worlds enclosed by seemingly impenetrable walls, freeing themselves from the sort of existential nightmare Sartre associates with the wall, but vibrating outward, to adopt the words of Gibbons, and like the ceramics employed by Koolhaas, telling the story of the life that takes place inside as a generational epic.

Brufe社区中心/Imago
S住宅/Suga Atelier
日光住宅/Takeshi Hosaka建筑事务所
Mecenat美术馆/Naf建筑设计事务所
Moliner住宅/Alberto Campo Baeza
内在丰富性的探索/Diego Terna

内在丰富性的探索

一面似乎不可渗透的墙壁

"……听,他们要把我们带到院子里去,嗯,接着他们会在我们面前站成一排,他们有多少人?"

"我不知道。五个或者八个吧。不会有太多。"

"好吧,就算他们有8个人。有人会喊:'瞄准!'接着我就能看见8支步枪正对着我。我想那时候我一定特别想往墙里钻,我会使尽全身力气用脊背顶住那面墙……,但墙岿然不动,就像在噩梦中一样。所有一切,我都想象得出来。你真不知道这一幕在我眼前有多么逼真!"——让•保罗•萨特,《墙》,1939年

白天,内部

埃托雷•斯科拉于1987年执导了影片《家庭》,Carlo是该片的主人公,他慢慢地走过一栋罗马公寓的长廊,这是一个独特的地方,房间都通向走廊,这里也是影片故事结尾的地方。Carlo(由维托瑞欧扮演)整理了一下自己的领带,进入了一个房间,在这里他拍摄了一张照片,照片里包含了在这栋房子中住过的好几代人的物品。

这部影片讲述了照片中这个家庭长达80年的历史——从上世纪初期直到上世纪80年代——影片完全是在这栋罗马公寓的室内拍摄的,浓缩了生活中的种种喜怒哀乐,在有限的空间内,各种角色如走马灯一般变化,却并没有让人觉得墙带来了约束感,人们反而能在这些墙之间找到与他们命运的共鸣。

在这种情况下,家庭成了人类行为典范的发动机,需要有实际上的限制以更好地表达成员的复杂潜能,在狭窄的空间中寻找潜能的爆发点,在他们有限的活动空间中几乎就可以寻求安慰。

首先,家是一个非常私密的地方,这里的生活很简单,没有陌生人窥探的目光:没有繁文缛节,也没有社会行为——或者更确切地说,社会性非常有限,也正因为有限,所以才没那么让人压抑。

也许是因为这个原因,家庭生活中最有趣的一些例子才可以在古老的庞贝遗址中寻得蛛丝马迹。这座古城本身就处在一个非常有限的生活空间里(同现在的许多城市空间相比),在不断地加建中形成,从外表看非常宁静,但是内部却人口稠密,热闹非凡。

事实上,房屋的内部结构是罗马人的发明,它非常新颖,正如建筑者布鲁诺•塞维在1997年所说的:古埃及、苏美尔文明、克里特岛的迈锡尼文明世界、伊特鲁里亚人和希腊人都忽略了这个空间,这个经验丰富、意义重大、或扩张或收缩的空间,传达出了特殊的建筑信

Brufe Social Center/Imago
House S/Suga Atelier
Daylight House/Takeshi Hosaka Architects
Mecenat Art Museum/Naf Architect & Design
Moliner House/Alberto Campo Baeza
Introversion, or about Observation (within)/Diego Terna

Introversion, or about Observation (within)

A wall, which seems impermeable

"[…] Listen, they're going to take us into the courtyard. Good. They're going to stand up in front of us. How many?"

"I don't know. Five or eight. Not more."

"All right. There'll be eight. Someone'll holler 'aim!' and I'll see eight rifles looking at me. I'll think how I'd like to get inside the wall. I'll push against it with my back…. with every ounce of strength I have, but the wall will stay, like in a nightmare. I can imagine all that. If you only knew how well I can imagine it." - Jean-Paul Sartre, *The wall*, 1939.

Interior, day

Carlo, the protagonist of the movie *The Family*, by Ettore Scola, 1987, slowly covers the long corridor of a Roman apartment, a unique place, along with the rooms that open on the corridor, in which the entire story of the film plays out. Carlo, played by Vittorio Gassman, adjusts his tie and goes into a room, where he will be taking a photo that contains all the components of the many generations the house has welcomed.

The movie, which tells eighty years of the history of the family portrayed in the photos — from the beginning of the 1900s until the late Eighties — is shot entirely in the interiors of the Roman flat, a microcosm that exhausts all the possibilities of a complex life, that does not feel the physical constraints set by the walls surrounding the changing cast of characters, but that finds, between these walls, a sounding board for their fates.

The family, in this case, becomes an engine of exemplary human action, needing physical limits to better express the members' own complex potential, and finding in the physical compression an acceleration of the same, almost taking reassurance from the limited size in which they act.

The home becomes, above all, a place of discreet privacy, where the living is easy, free from the gaze of strangers: etiquette doesn't exist there, then, nor does social behavior — or rather, there is a limited sociality, and because limited, possibly less inhibited.

Perhaps for this reason some of the most interesting examples of family living are found in the ruins of ancient Pompeii, where the city, itself compressed within a limited space (as compared to many current urban spaces), builds itself by adding houses which are fairly mute from the outside, but densely concentrated and lively on the inside.

The interiors, as a matter of fact, constitute a real novelty as a Roman invention, as related by Bruno Zevi in 1997: *Ancient Egypt, the Sumerian civilization, the Cretan-Mycenaean world, the Etruscans and Greeks ignore the space, the experienced, meaningful, dilated or contract voids, which convey the specific message of architecture;*

拥有巨大城墙和零星洞口的斯卡利杰罗城堡

Scaligero Castle with huge walls and few openings

庞贝古城"威第之家"中的天井

atrium of the "House of Vettii" in Pompeii

息；然后继续引用李格尔（20世纪初奥地利著名艺术史家）的话：借助墙体才能实现到现在为止还被认为是不可能的事情，那就是空间的可识别性。

因此古罗马的建筑第一次表明，建筑的用途与雕塑或景观并没有太多关联，它为内部腔体留出了体量，为建筑表皮内的封闭空间预留出了内容。从这层意义上说，庞贝古城采用了一些经过深层次挖掘和慎重思考而得出的内向性实例：一个是威第之家，始建于公元前一世纪，在公元一世纪又进行了改造。

在庞贝古城，公元79年发生的大规模火山爆发将这里的现实生活永远定格在了那一瞬间，也许这里就是处理内部空间的第一种方案的诞生地。在这种方案中我们看到了第一个社会核心的标志——家庭，这并不足为奇。接着，家庭空间成了展现对建筑的全新看法的地方，人们能够在其中感受到不同寻常的空间感。

威第之家是一个传说，它沿着许多巨大的腔洞往前移动，一个接着一个，形成了错综复杂的路线，屋顶上的方形大洞能让光线照射进来。以这种方式，王宫一般的宏伟壮观再一次呈现在普通民居之中，从而构建出了丰富内部空间的不断变换的景观。因此房屋和城市之间并没有展开对话，而众多的封闭空间之间却联系紧密，例如在"腔洞"中环绕着外部空间的房间，像做实验一样被包围在住宅的墙壁里面。

于是，居住空间变得与公共职能息息相关，但是却专门为建筑物的使用者服务：经过精挑细选的小镇改善了几大主要特色（绿草如茵、符合标准的光线、河道以及调控得宜的温暖感觉），最后给人感觉仿佛是驯化了的野生动物。

因此，家庭成员并不用寻求向外扩张：他们已经在广场上、市场上、教堂中参与到了公共生活之中。在他们自己的私人生活范围中，他们感受到一种可控的生活状态，在城市的两种生活形式之间实现了有益的分离。

夜晚，外部

2012年6月26日，Portishead乐队在意大利的一栋1202年竣工的中世纪建筑——位于维罗纳附近的维拉法兰卡的斯卡利杰罗城堡——中举行了演唱会，这是他们此次欧洲巡演的两个意大利站中的第一站。

高大的锯齿形城墙环绕着一个宽阔的绿色庭院，这个地方很明显与环绕着它的小城市比例失衡，因而能够使这个场所成为城市空间感的核心位置。

不需要透明和繁复的通道：内部庭院与城市之间的关系沿着高高

then to continue quoting Alois Riegl: *with what, was achieved what until now was considered impossible, namely the identification of the space.*

The works of ancient Rome thus showed, for the first time, the use of architecture for purposes that had little to do with sculpture or landscape, giving volume to internal cavities and substance to the enclosed spaces within the physical limits of the buildings. Pompeii, in this sense, introduces examples of a deep exploration of deliberate introversion: one such is the House of Vettii, originally built in the first century BC and then modified in the first century AD.

Here, where the great eruption of 79AD has crystallized forever a moment of real life, seems to be revealed the first approaches to the interior space. Not surprisingly, we can observe in this approach the mark of the first social nucleus, the family. The domestic spaces, then, become places for exercising a new attitude toward architecture, as territories in which one can experience an uncommon spatiality.

The House of Vettii is a tale that moves along sizable cavities, which follow one another to create complex and composite paths, piercing the roof and admitting light indoors. In this way the large scale of imperial palaces is again presented on the domestic scale, allowing the construction of ever-changing views that enrich the interior. The dialogue, therefore, is not between the house and the city, but between the proliferation of closed spaces, such as rooms which surround outdoor spaces in "captivity," encased, as an experiment, within the walls of the house.

The domestic space then becomes a pulsation of public functions, but dedicated exclusively to users of the building: miniature towns that are selected, exalted in their main features (the lush greenery, the calibrated light, the channelled water, the regulated warmth) and finally, as if they were elements of a sort of wildlife, domesticated.

It is not necessary, therefore, for family members to seek to expand outwards: they already participate in public life in the fora, in the markets, in the basilicas. Within the volumes of their own personal living, they experience a controlled life which allows a beneficial separation between the two forms of life in the city.

Exterior, night

On 26 June 2012, Portishead played the first of two Italian dates on their European tour. They performed in a medieval building, completed in 1202: the Scaligero Castle, in Villafranca, near Verona.

Here the high crenellated walls enclose a large green court, a place clearly out-of-scale with the small city built around it and thereby capable of centralizing the urban spatiality of the place.

OMA建筑事务所于2005年设计的位于波尔多的音乐之家，项目为封闭的多面体形式，由大块坚固的白色混凝土制成

Casa da Mùsica in Porto by OMA, 2005 with a closed faceted form made of solid white concrete mass

的砖墙在回荡，嵌入城垛之间，又与天空融合在一起。

正是在这里，就在观众们大喊着"再来一个"之前，贝丝•吉本在歌曲Threads的最后几句发出了歇斯底里的呐喊，维罗纳城中斯卡利杰罗城堡的墙壁在这个看似娇弱的女人迸发出的爆发力下仿佛要迸裂一般。

因此，建筑物乍一看似乎给人一种内向的感觉——高墙环绕的庭院——显示出它如何将自身变成了与户外相通的构造：墙壁将不再把内外分隔开来，而是一种振动膜，哪怕只有一丝声响，也会反映出在内部绽放的浓缩人生。

它们的确是相同的墙，让人们的感受在这里凝聚，让他们随着这支英国乐队一起放声唱歌。也许只有到那时，声音才能四散而去，绚烂的色彩也可以从舞台上发散，遁入大街小巷，最终都落到了那些好奇的新观众的眼睛和耳朵里。

此外，位于波尔图的音乐之家是OMA建筑事务所于2005年实施的项目。该建筑坐落在地面，就像一个天外来客一样，似乎与环绕它周围的城区毫无关系。任何空间上的参照物都是倾斜、扭曲、失真的，指向天空的墙面各个倾斜角度都有，土壤也似乎在这个庞然大物的重压之下变得弯曲了。

灰色的混凝土饰面为皮拉内西（意大利建筑师、艺术家）风格的观景建筑提供了一种表现形式，在这里梁（或柱：它们均无法定义）、楼梯、墙壁与地面交错纵横，在这些似乎大到无边无际的体量中扭曲变形。同样，大面积的窗户也似乎无法与城市展开对话，但是人们却将它们简单地定义为不同于混凝土的材料，仍然能够呈现出与混凝土相同的庞大的存在感。

然而尽管拥有这些初步的观察，建筑物仍然是一个杰作，就算没有面对整座城市，它也会带给整个周边地区一种存在感：正如在维罗纳，城堡的墙壁与内部声音一起颤动，两者形成共振直达城镇当中，所以库哈斯在波尔图设计的这栋建筑能够作为一种膜，呈现出建筑物内部生活的音乐流动形式。这位著名的荷兰设计师用几个开口完成了这栋建筑的设计，使人们可以沿着楼梯退回到地面，仿佛该建筑结构固定在外星星球上一样，但是最重要的是，有一种氛围从每个不对称设计的角落中散发出来，将这个庞然大物转变为与城市对话的构造。

在威第之家中，由众多天井构建了家居空间，在私密的空间里重新唤起人们对公共生活的记忆，在这里，一连串小房间将城市强势地带到了建筑物之内：这些房间大小有限，陶瓷装饰立即唤起了人们对于本土特有材料的回忆。由于使用了代夫特陶器，这些元素使参考标

No need for transparency and frequent passages: the relationship between the inner court and the outer city resounds along the high brick walls; it wedges itself between the battlements; it escapes into the sky.

It was here that Beth Gibbons burst into her extreme and furious scream during the final bars of *Threads*, just before the encore; the walls of the castle seemed to explode with the force springing from this deceptively fragile-seeming woman.

So, what at first sight might seem to be a feeling of introversion — the court, enclosed by high walls — showed how it can transform itself into a mechanism for approaching the outside: the walls are no longer a division, but vibrating membranes that return, even with a hint of sound, the concentration of life unfolding inside. They are, indeed, the same walls which allowed the feelings of the people gathered here to condense, as they sang along with the English group. Only then, perhaps, could the sound escape, as well as the colors emanating from the stage, to lie down in the streets, calling out, finally, to new curious spectators.

Also, in Porto, lies the Casa da Mùsica, opened in 2005 as a project of OMA. It rests on the ground like an alien object, with seemingly nothing to do with the urban area surrounding it. Any dimensional reference is tilted, twisted, distorted: walls point crookedly to the sky, and the soil itself seems bent under the weight of the gigantic object.

The concrete gray finish gives form to the container of Piranesi's views, in which beams (or pillars: they are impossible to define), stairs, walls, floors intersect, twisting within volumes that seem immense. Similarly, the large windows seem unable to enter into a dialogue with the city, but are defined simply as squares of a different material from the concrete, while still achieving the same massive presence.

Yet despite these preliminary observations, the building is a masterpiece that manages to give a sense to the entire neighborhood, if not to the whole city: as in Verona, where the castle walls vibrated with the sound inside, resonating it into the town, so in Porto this building by Koolhaas acts as a membrane that reveals the musical flow of the life in its interior. The Dutch architect accomplishes this with a few openings that allow the stairs to exit to the ground, mooring the structure as on an extraterrestrial planet, but, above all, with a sort of aura that emanates from every lopsided corner and that transforms the great object into a mechanism of civic dialogue.

Where in the House of Vettii the patios structure the domestic spaces, recalling public life in intimate spaces, here a series of small rooms bring the city strongly into the building: they are rooms of limited size, lined with ceramics which immediately re-

S住宅，运用的纯粹形式让人想起深深的洞穴

House S, the pure form reminds of a deep cave

在建筑表皮上有切口的Brufe社区中心，能够抵达建筑物的中心位置

Brufe Social Center with the cuts on the skin reaching the heart of the building

准变得更加复杂，因此将库哈斯的故乡荷兰与葡萄牙及其花砖相结合，并未区分前后二者的差异。

但是在建筑物的表面上已经建立起了与城市的真正对话。在屋顶的倾斜表面上有一个切口，它是展现在建筑物生动外观上的切口，因而形成的一个小庭院呈现出了可贵的简洁而丰富多变的表面，城市的光线投射在这些表面上，在下面都无法看到，然而它们在顶部却占据着主导地位。

回到建筑物内部

这里收集的一些项目展现了建筑对城市环境的态度，这与我们之前看到的非常相似：一种定义为内向性的态度，而事实上呈现出的远远不只是在划定的界限内简单地被关闭的状态。

在某种程度上，每个项目都是以观察中的事态为基础：第一，观察自己，自己的家庭，那是封闭在建筑物中的小社会；接着观察外部，反过来被观察，在这个对建筑物与城市进行装点的观察游戏中彼此不断渗透。

由Suga Atelier设计的S住宅以及由Naf建筑设计事务所设计的Mecenat美术馆，通过天井系统在家庭中定义了一个开放式空间，就像庞贝古城的威第之家一样。在这里墙壁的外表面被整合成一个宽阔的露台，自然意味着外部的光线和视野，似乎要重新构建内部的空间，从而营造出多个景观，使视线能够穿过复杂的景象。

因此，居民并不会有一种被包围的感觉，而会感觉到社交活动有多种不同的可能性：在这些区域内，人们会涉及与外部之间的关系，在炎热与寒冷中，在光明与黑暗中，所有的一切都在可控的状态里。住宅变成了一个都市化的项目，由小路、虚实空间、广场及街道组成。

相似的情况在埃托雷·斯科拉的影片中也有所体现，住宅的墙壁成为故事发生的背景，虽然规模较小但却让人更加熟悉：走廊就是街道，将每栋建筑体量中的空间连接在一起，诉说着人性的千姿百态。

由Imago设计的Brufe社区中心环绕在内部空隙周围，为生活空间预留了尺度，但是与外部的关系却少了一分冷硬，多了一丝细致入微。正如在OMA的项目中，Imago也运用高低不平的饰面来打造建筑物，但是在这里，结果似乎没那么神秘莫测。形成这种差异的原因在很大程度上是建筑表皮上的开口形成的观景视野所造成的，开口能够抵达建筑物的中心位置；随着景观不断扩大，内外空间的界限变得模糊，轮廓鲜明的界限消失了。接着，不断增加的渗透性削弱了建筑物的庞大体量与厚重感，通过对复杂性的解读来提升建筑物的品质，为它蒙上一层面纱，却激发了人们更强烈的好奇心。

call the characteristic materials of the nation. These manage to complicate the references, thanks to the use of Delft pottery, thus combining the Netherlands, home of Koolhaas, with Portugal and its azulejos, without distinguishing the former from the latter.

But it is in the building's covering that a true dialogue with the city is established, thanks to a tiny courtyard, the result of a cut in the sloping surface of the roof, an incision into the lively skin of the building which reveals the preciousness, the simplicity, of a chequered cover, all projected toward the lights of the city, invisible from below, yet dominant on the top.

Back to the inside

The projects collected here show an attitude towards urban surroundings very similar to what we have seen previously: an attitude which is defiantly introversion, but which actually reveals far more than a state of simply being shut within circumscribed limits.

In a way, each of these projects bases its affairs in observation: first, to observe oneself, one's own family, the small society enclosed in the building; then to observe the outside, being observed in turn, in a game of glances which invests the building and the city, interpenetrating.

The House S by Suga Atelier and the *Mecenat Art Museum* by Naf Architect & Design, define an open area within the home, as happened in Pompeii's House of Vettii, through a system of patios. Here the enclosure of the walls is dissolved into a wide patio, where nature, meant as light and vision of the exterior, seems to retake possession of the interior, creating multiple landscapes in which the gaze can move across complex sights.

The feeling for the inhabitants, therefore, is not one of enclosure, but of a different possibility for sociality: in these areas one becomes involved in the relationship with the outside, in hot and cold, in light and darkness, all in a controlled manner. The house becomes an urbanistic project formed of paths, of full and empty, of squares and streets.

Something similar happens in Ettore Scola's film, in which the walls of the house become the background of stories now public, albeit on a smaller and more familiar scale: the corridor is the street, which connects spaces in which each volume speaks of different human characteristics.

Brufe Social Center designed by Imago also organizes itself around internal voids that give the measure of living, but in which the relationship with the outside is less hard, more nuanced. As in the OMA project, Imago also works with a rugged finish, but here the result seems less enigmatic. This difference is due largely to the views created by cuts in the skin that reach the heart of the building; the views expand, the internal and external spaces acquire

图片提供：courtesy of Alberto Campo Baeza(©Javier Callejas)

Moliner住宅，周围环绕着白色的墙壁，光线透过大面积的半透明玻璃散射到室内

Moliner House, surrounded by white walls with light diffused through large translucent glass

图片提供：Takeshi Hosaka建筑事务所
(©Koji Fujii/Nacasa&Pertners公司)

日光住宅，日光可直射到弯曲的丙烯酸天花板下面，这个天花板也是除建筑主门外的唯一开口

Daylight House with direct lights down the curved acrylic ceiling, the only opening except the main door of the building

同样，Alberto Campo Baeza设计的Moliner住宅呈现出一种更简单的系统，但是这种最低限度的简单却获得了绝对的价值：住宅位于建造地点的中心位置，周围环绕着一面墙，墙壁限制了建筑物的属性，同时也隐藏了建筑物的大部分玻璃立面。庞贝住宅的复杂性毫不显眼，建筑姿态脱离了建筑物的外观，从而在住宅的墙壁和体量中呈现出一种对立关系。从这个意义上说，该项目似乎采用了与斯卡利杰罗城堡相似的处理方法：在这个案例中以及被围墙包围的周边区域都失去了其实质性，住宅大楼向上耸立预示着在建筑内部举行的活动拥有很高的私密性。接着，明显的内向性在这里爆发，也许没有通过音乐，而是通过体量的对比。

最后的实例比较极端：Takeshi Hosaka建筑事务所设计的日光住宅消除所有与外部接触的机会，房子上没有开口，没有窗户；住宅的墙壁是完全不透明的。然而一进入这栋住宅，人们就会感到意想不到的明亮，那是一种令人惊讶的微光。事实上，建筑表面包含天窗系统，会让内部受到大量的光照。在这个从庞贝古城开始扩展的主题中，在斯科拉影片中的全家福中找到的表达方式中，Takeshi Hosaka表现了一种需要集中注意力的生活方式，以人和光线之间的主要关系为基础，光线在这里作为一种建筑材料在所有的建筑意图中均有所呈现。因此，房子中的居民失去了与垂直墙壁的联系，只能将他们自己与向上的抽象词语相比较，寻求一种返璞归真的生活方式，就像以前的人类活动都是根据天色决定的。

并非不可渗透的墙壁

萨特在墙壁上发现了一个让人无法逃离可怕现实的限制条件，一个可能无法穿越的对象，并对来自受害者的要求置若罔闻。

然而在这里我们看到，在现实中，内外之间的关系并不会因墙的存在而结束，但不知何故人们却又强调这种理念，将它带入城市当中，带入居住者的行为之中，仿佛墙会渐渐地变薄，允许在两个明显分开的世界中互相渗透。

贝丝·吉本的尖叫声仍然让人们记忆犹新，这声音渗透到了中世纪城堡的墙壁中，同时也例证了在住宅和城市之间，内敛性是如何自我转变为唤起人们回忆的过滤器的。Diego Terna

ambiguities, and sharp limits disappear. The massive strength of the building, then, is tempered by an increased permeability, enhancing the architecture with a note of complexity, making it less readable, though arousing greater curiosity.

Similarly, the *Moliner House* by Alberto Campo Baeza evinces a simpler system, but this minimal simplicity acquires a value of absoluteness: the house stands at the center of the lot, surrounded by a wall that restricts the property and hides the primarily glass facades of the building. The complexity of the Pompeii house is nowhere in evidence; the flesh is taken off the architectural gesture to reveal a counterpoint between the wall and the volumes of the house. In this sense, the project seems to take an approach similar to that of Scaligero Castle: in this case as well the walled perimeter loses its materiality, with a house–tower that rises upward foreshadowing the events that take place inside, which enjoy a high degree of privacy. The apparent introversion, then, explodes here as well, not through music perhaps, but via contrasts of volume.

The final example is the most radical: The *Daylight House* by Takeshi Hosaka Architects eliminates any confrontation with the outside; there are no openings, no windows; the walls of the house are completely opaque. However, entering the house, one discovers unexpected brightness, an astounding glimmer. The cover, in fact, consists of a system of skylights which rain light upon the interior. In this expansion on a theme that began in Pompeii and found expression in Scola's family portrait, Takeshi Hosaka suggests a lifestyle that requires concentration, based on a primary relationship between humanity and light, which is here used for all intents and purposes as a building material. The inhabitants of the house, therefore, lose contact with the verticality of the walls, comparing themselves only with the abstract term of upward, seeking a kind of primitive living in which human affairs follow the luminous flux of time.

A wall: Not impermeable

Sartre found in the wall a limit that prevented escaping from a dire reality, an object that could not be crossed, deaf to the demands of the frightened condemned.

Yet here we see how, in reality, the massive presence of a wall does not end a relationship between outside and inside, but somehow exalts it, bringing to the city the actions of the people who live inside, as if, little by little, the wall thins itself, allowing a permeability between two apparently divided worlds.

The scream of Beth Gibbons is still a vivid memory permeating the walls of a medieval castle, an example of how an introverted character can transform itself into an evocative filter between the house and the city. Diego Terna

Brufe社区中心
Imago

内在丰富性建筑 | Introvert Potential

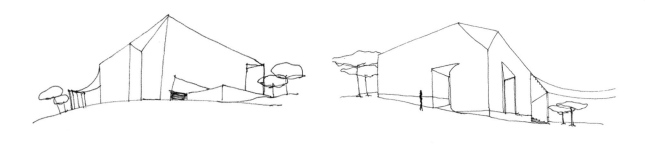

该设计方案所建造的建筑对社区至关重要,服务于葡萄牙法马利康新镇的Brufe村。该建筑拥有良好的社会秩序,在单一的空间内集多样性的服务于一身,从牙牙学语的孩童到年至耄耋的老人,可满足不同年龄段居民的需求。

该社区中心提供以下服务:可容纳20人的老年人日间照管中心、容纳33名儿童的托儿所以及容纳24名老年人的养老院。该中心也能够每日为30名老年居民提供上门护理服务。

该建筑采用削减式体块,切割了一些开口用来实现内部空间的照明,在大多数情况下,这些开口将建筑打开,露出了内部庭院。建筑将本属于内部的结构外置,而外立面厚重而不透光,只有像楼梯和正门入口这样的主要结构留有一些开口。室内立面看起来是通透而延续的玻璃幕墙,树荫遮住了由玻璃幕墙围合的内部露台,提供了安全可用的开放空间。在建筑主体的顶部,每个切口都衬托着周围的一种功能性设施,包括带顶停车区、正门入口和仿照古罗马竞技场用于举办室外活动的带顶看台区。

该项目设计的标准是将不同功能区围绕在中心庭院周围。其自身的交流加强了每个区域元素间的固定视觉关系。每个区域元素虽以一种固定的方式分布,但彼此之间既相互联系,又在必要的时候相互独立。Imago

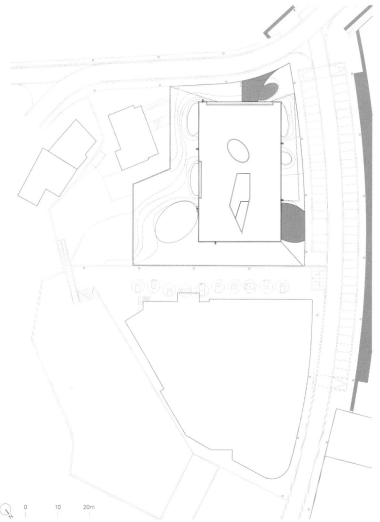

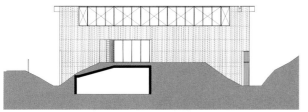
西南立面 south-west elevation

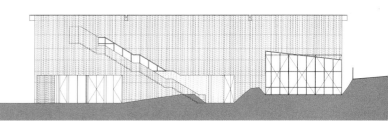
西北立面 north-west elevation

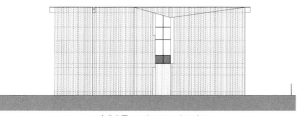
0 2 5m 东北立面 north-east elevation

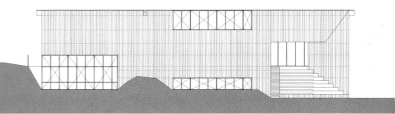
东南立面 south-east elevation

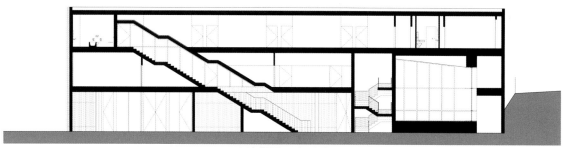

A-A' 剖面图 section A-A'

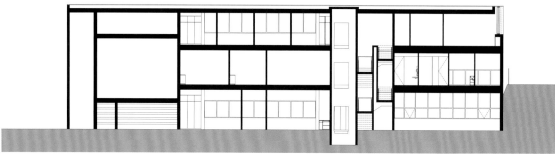

B-B' 剖面图 section B-B'

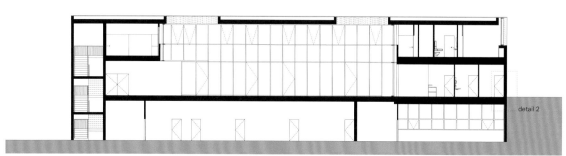

C-C' 剖面图 section C-C'

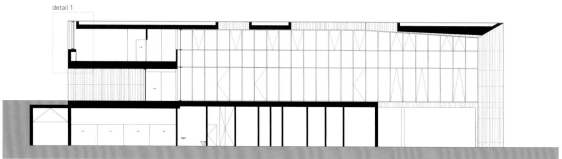

D-D' 剖面图 section D-D'

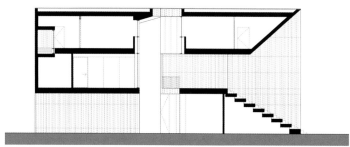

E-E' 剖面图 section E-E'

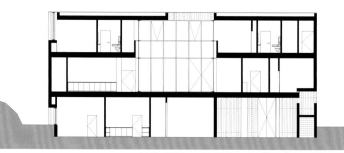

F-F' 剖面图 section F-F'

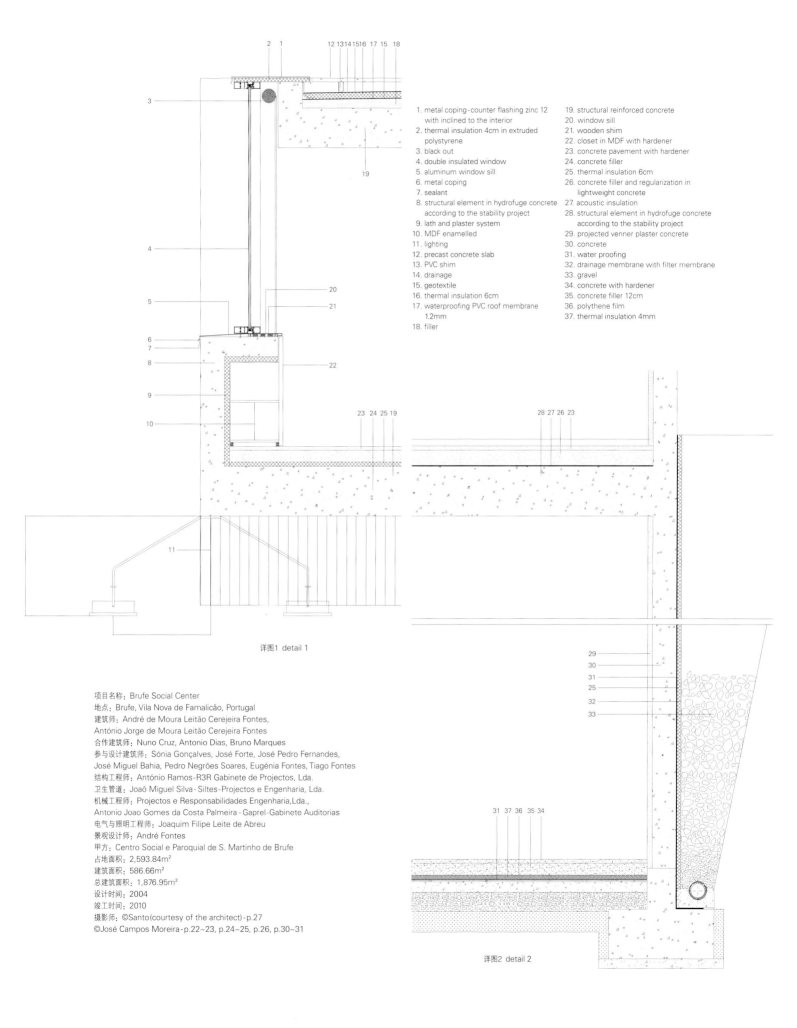

Brufe Social Center

The proposal was based on creating a building of fundamental importance for the community that would serve the parish of the village of Brufe, Vila Nova de Famalicão in Portugal. Possessing valences of social order, the building gives the population, in one single space, a variety of services that integrates all age groups, from very young to elderly.

The community center will consist of the following services; an elderly day center that will serve up to 20 people, a day care center and nursery that will serve 33 children and a rest home for 24 elderly residents. The building will also be able to provide extended care at the homes of 30 elderly citizens daily.

Based on a subtractive block form, some openings are carved out to illuminate the interior space or in most cases tearing and piercing the building to its inner courtyard. The building is turned inside out; the exterior facades look opaque, dense, with just a few openings in the main points of the building like the stairs and the main entrance. The interior facades appear as a translucent

and continuous glass curtain wall that embrace the internal patio shaded by proposed trees and providing a protected usable open space. On the top portion of the mass each carving provokes an event with its surroundings, covered parking spaces, main entrance, and a covered seating area for open air events simulating an amphitheater.

The programs criteria is organized in functional components that surround a central courtyard. The physical communication within itself accentuates a permanent visual relationship with each component. There can be an interconnection with each component or independence when necessary and distributed in a permanent manner. Imago

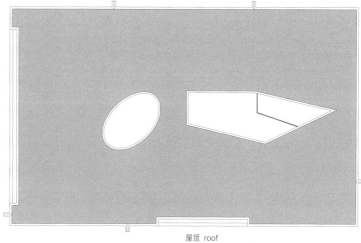

屋顶 roof

1 双人卧室（成对单人床）
2 单人房间
3 休息区
4 养老院
5 双人卧室（单人床）
6 休息室+小厨房

1. double bedroom (twin beds)
2. single room
3. rest area
4. rest home
5. double bedroom (single bed)
6. lounge room + kitchenette

二层 second floor

1 教研室
2 园长办公室
3 大厅/接待处
4 儿童更衣室
5 活动室（2-3岁）
6 活动室（1-2岁）
7 教室（0-1岁）
8 产房
9 库房
10 主管办公室
11 行政办公室
12 会议室
13 医生办公室
14 休息室
15 活动室
16 多功能室
17 室外露台

1. teacher's office
2. director's office
3. hall/reception
4. children's dressroom
5. activities room (2~3years)
6. activities room (1~2years)
7. educator room (0~1years)
8. maternity ward
9. storage room
10. direction office
11. administration office
12. meeting office
13. doctor office
14. lounge
15. activities room
16. multidisciplinary room
17. exterior patio

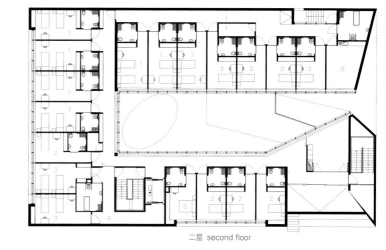

一层 first floor

1 儿童食堂
2 儿童户外活动空间
3 养老院食堂
4 小厨房
5 库房
6 厨房
7 洗衣房
8 员工休息室
9 露天礼堂
10 设备间

1. children's refectory
2. children's exterior space
3. rest home refectory
4. kitchenette
5. storage room
6. kitchen
7. laundry
8. employee lounge
9. open air auditorium
10. technical room

0 2 5m

地下一层 first floor below ground

S住宅
Suga Atelier

有些事物人们很难发觉其存在，除非我们找到合适的语言来形容它。

在一条主干道旁就坐落着这样一座建筑，它丝毫不受周围来往车辆的影响。该建筑使用结构胶合板，凹凸不平的表面形成其框架，整体看起来像粗糙的混凝土挡土墙。从东面两侧墙体掩护的裂缝处——就像防御所用的射击孔——向外望去，可以欣赏到日出和整座城市的景色。

然而，当入口大门这一主要的光线来源被切断时，整个空间也随之封闭起来。但建筑中间的天窗使整个空间依然保持向空中开放的状态，而入口的位置使得天窗呈现不规则的L形。空间内部布满回收再利用的塑料瓶，与胶合板相同尺寸（910mm×120mm）的白色聚酯纤维垫子，用隔板与塑料螺丝钉固定在墙体和天花板上。

地板上铺设了以相同材料制作的针刺无纺布地垫。除具有保温隔音作用外，这一设计还体现出整个密闭空间易于吸收光照的特色。整个空间显得十分静谧，犹如深邃的洞穴。走在上面也仿佛走在榻榻米垫上，有弹性的墙体就像一层缓冲垫。人们可以随意坐卧在建筑内的地板上，该设计是对坐在地板上这一新潮生活方式的尝试。另一方面，像楼梯和阁楼这样与建筑相关的家具设施都使用了回收再利用的胶合板，留住以往的回忆。从天空投射进来的光线洒在白色的地板上，通过始终位于正面的九边形玻璃聚光，形成了一幅光影闪烁的画面。雨滴穿过采光天井，落在位于车库层的浅水池中，散开层层涟漪。

Suga Atelier

项目名称：House S
地点：Osaka, Japan
建筑师：Suga Shotaro
结构工程师：Fukunaga Takeshi
机电工程师：Ikeda Yoshihiko
照明工程师：Suga Shotaro
景观设计师：Suga Shotaro
占地面积：89.35m²
建筑面积：53.59m²
总建筑面积：115.31m²
竣工时间：2012
摄影师：courtesy of the architect-p.32, p.34~35, p.36, p.37, p.38, p.39, p.41, p.44, p.45
©Yuko Tada(courtesy of the architect)-p.33, p.40, p.43

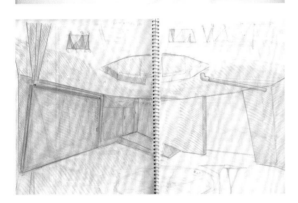

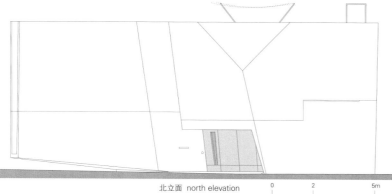
北立面 north elevation

东立面 east elevation

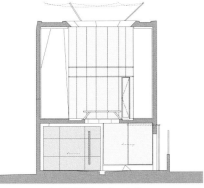

A-A' 剖面图 section A-A'

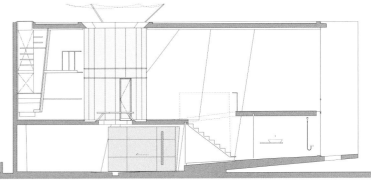

B-B' 剖面图 section B-B'

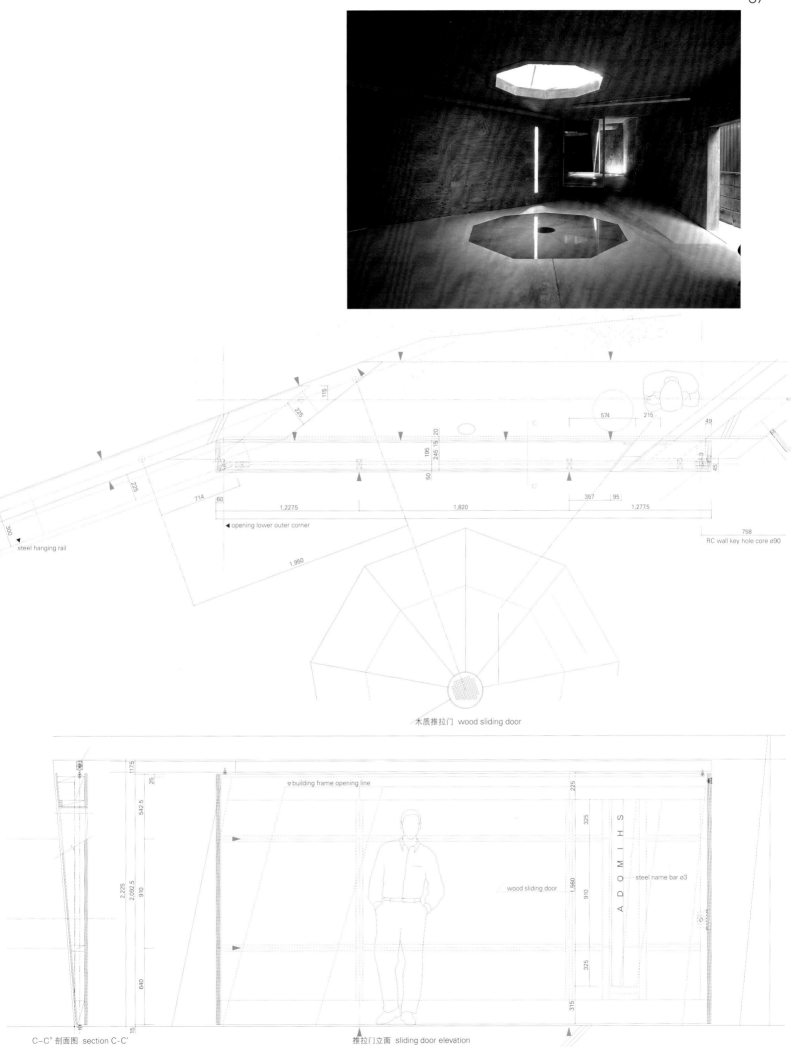

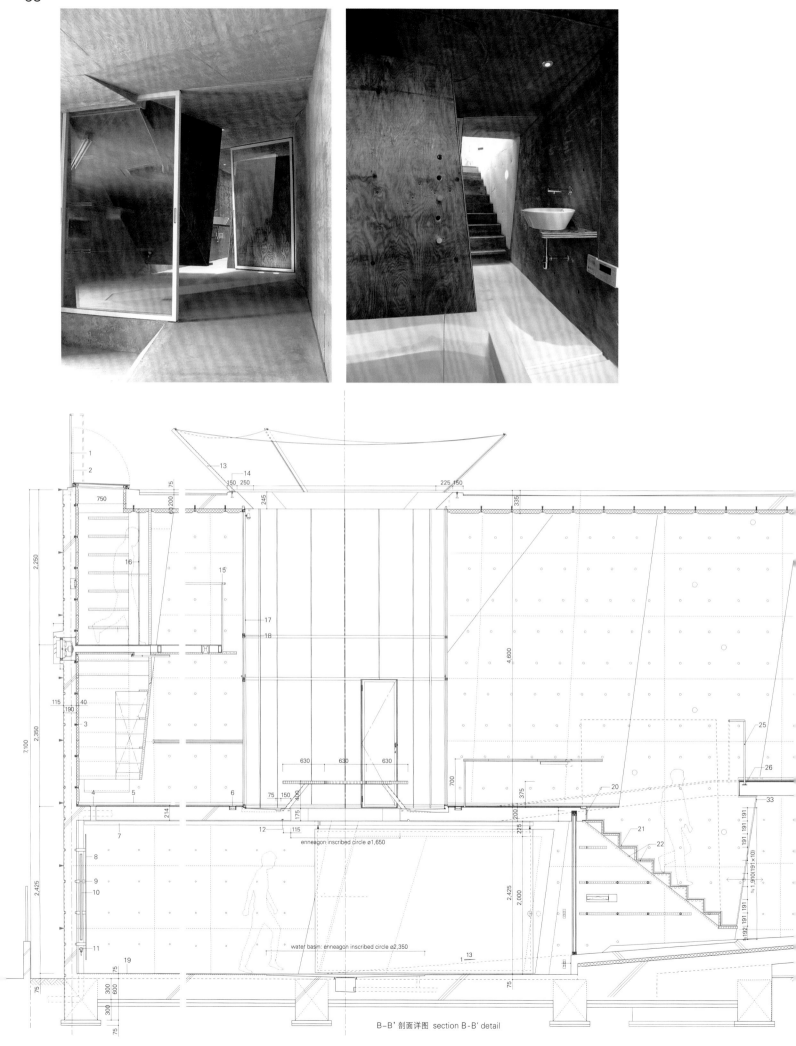

B-B' 剖面详图 section B-B' detail

1. handrail: ST ø25 galvanizing
2. SS hinge with damper
3. plastic screw
4. floor opening: TG t8 ø100
5. base: PP joiner U shape 25
6. recessed luminaire: TG t8 ø100
7. steel hanging rail
8. cover ST-PL t4.5 910×1,820
9. fixation: long nut×8
10. seamless line
11. faucet
12. throating joint 20×15
13. tent support: ST L50×50 galvanizing
14. SS insert M12
15. handrail: ST ø25
16. hangs material: ST ø19
17. mullion ST L 50×50
18. door frame: ST L40×40
19. floor: V cut water works t10
20. rough concrete
21. joiner: U shape 15
22. reuse plywood
23. waterproofing rise
24. drain hole VP ø25 opening
25. handrail+support: ST ø25 bending
26. embedded welding PL
27. white cement
28. heat insulating materials t30, t50
29. RC slab t200+waterproofing
30. polyester fiber insulation mat×2
31. direct control+dustproof paint
32. plastic joiner: U shape 25
33. joiner
34. ceiling insert
35. curtain rail embedding
36. boundary of street level
37. soil with clover seeds
38. direct control slope
39. coat waterproofing
40. RC slab t200
41. ceiling insert: W5/16
42. insulation: polyester fiber insulation mat t60 plastic screw

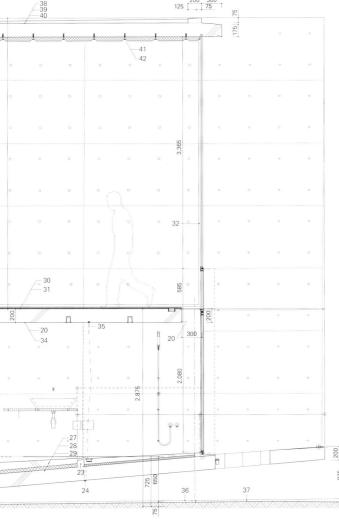

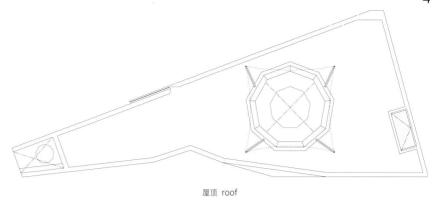
屋顶 roof

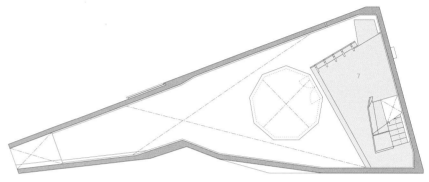
三层 third floor

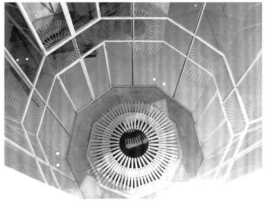

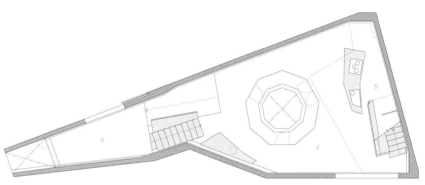
二层 second floor

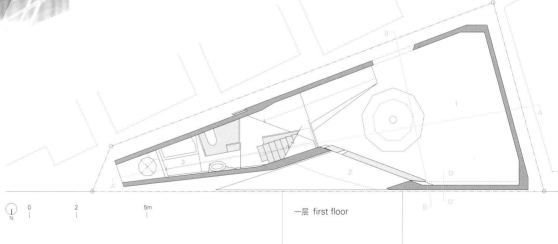

一层 first floor

1	车库	1. garage
2	入口	2. entrance
3	浴室	3. bathroom
4	起居室/餐厅	4. living/dining room
5	厨房	5. kitchen
6	夹层	6. mezzanine
7	阁楼	7. loft

House S

There are things that are hard to see even its existence, unless we find suitable words or expressions.

Beside a major road, the house is present, resisting against traffic. Since the house uses structural plywood that has pits surfaces as a formwork, the whole looks like a rough concrete retaining wall. When looking through a slit that is protected by walls on both sides like an embrasure on the east side where the site extends, morning sun and the city could be viewed. However, the space is closed when the large entrance door that is the only major aperture is shut down. Yet, the space is always wide open to the sky by the light court positioning in the center of the space that is slightly deformed in the shape of "L" due to the entrance. The interior of the space is covered by recycled plastic bottles, white polyester fiber mat in the same size of plywood (910mm×120mm), fixed to the walls and the ceilings with plastic screws using separators. Needle-punched mats made with the same material cover the floor. The idea was to express the light absorbing shining cocoon in addition to insulation and soundproofing. The space became very quiet and pure like a deep cave. The sense of walking onto it is like tatami-mat and the elastic walls are almost like a cushion. It is a trial of a new floor-sitting life style that offers sitting and lounging wherever in the house. On the other hand, furniture-like staircase and the loft, which are made relating the house, are made with recycled plywood used for formwork and keep the memoir. The light brought from the sky draws blinking picture of light on the white floor through enneagon glass that always maintains a positive position. Rain goes through the light court, beat down on the thin water basin on a garage floor and remain its traces. *Suga Atelier*

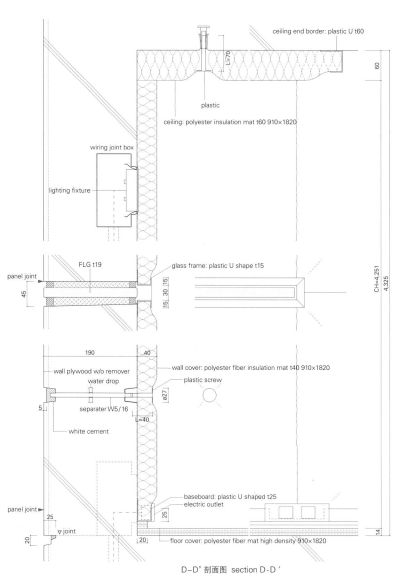

D-D' 剖面图 section D-D'

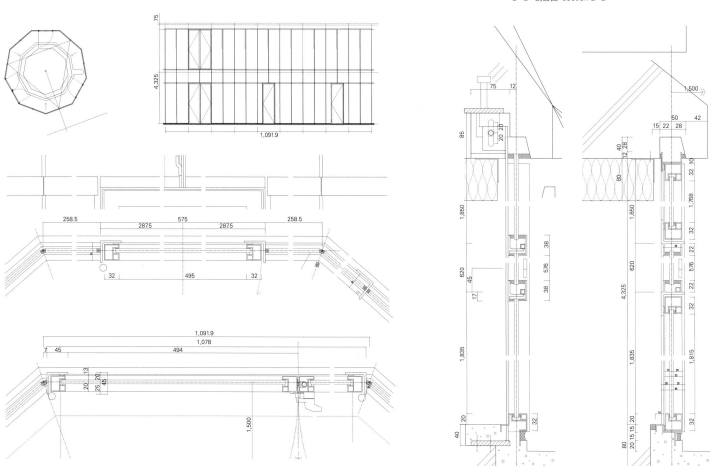

采光庭院遮阳板详图 light court screen detail

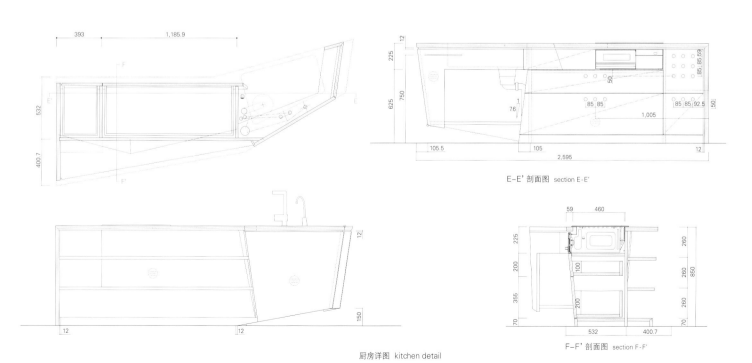

厨房详图　kitchen detail

E-E' 剖面图　section E-E'

F-F' 剖面图　section F-F'

内在丰富性建筑 Introvert Potential

日光住宅
Takeshi Hosaka Architects

在这座住宅里,居民生活在直接照射进来的自然光之下。

从该建筑步行至火车站仅需五分钟,其周围分布着独立的居民楼、十层高的公寓楼和一些办公大厦。由于该建筑坐落于高楼之间的低凹处,因此能获得自然光线就显得格外珍贵。一个四口之家打算在这里安家。

该建筑将一种单元网格(150cm×160cm)覆盖在整个基地上,使用挑高天花板的房间作为卧室、儿童房和书房,并利用家具将各区域分隔开,这些家具几乎占据了天花板高度的一半。从每个房间都能感受到整个屋顶的巨大跨度。

由于该住宅采用了透明的天窗,周围高楼里面的人们可以轻易瞥见该住宅内的一切,因此从一开始,建筑师就毫不犹豫地决定将天花板设置在天窗下面。通过安装天花板,人们可以准确地感受到天空微妙的变化和情况,无需抬头仰望。内部空间的布局如何使业主充分感受到室外的情况,就成了建筑师关注的焦点。天花板采用了白色亚克力材质的拱顶,自然光线可以透过天花板表面的玻璃天窗柔和地投射到屋内。除了灰泥浆地面和落叶松胶合板墙体,可视性高的混凝土白色亚克力表面经过了精细的处理,人们已经完全感觉不到混凝土的存在,仿佛没有使用砖瓦。这样,墙体表面失去了混凝土的存在感,变得更为抽象,仿佛只有光线才存在于建筑室内的上方一样。

每个房间都有一扇窗户(90cm×70cm)。透过这扇窗户,微风可以吹遍整座住宅,人们会感觉十分舒适。

建筑师在亚克力表面和屋顶之间留有空隙,运用加压气流原理,在夏天,它可以将阳光照射的热空气排出室外,而在冬天,又可使气流停止运动,并利用该空隙作为热量缓冲区,以确保室内温度稳定。

一进入建筑内部,就能感受到来自天空的光照,让人难以置信自己处在高楼环绕的低凹处。这座建筑被命名为"日光住宅"。日光的意思不仅仅指代太阳所散发的光芒,还代表着一整天里各种各样的美丽光线。比如说,在清晨黎明时分,可以看到天空颜色的变幻,或是湛蓝的天空,还有阳光直射时分,云层遮住太阳的影子,以及日落黄昏时天空的颜色,还有月光等。

晚上关灯休息时,透过天窗即可感受到夜空。该建筑天窗上折射的光在24小时内千变万化,给业主创造了一种即使在室内也能感受到外界大自然变化的空间氛围。

Daylight House

This is a house in which residents live under natural light from the sky.

The site is five minutes' walk from the railway station, and it is surrounded by a mixture of detached dwellings and 10-floor condominiums and office buildings. In this location nested in a valley between buildings, the light streaming down from the sky above felt precious. A couple with two children planned to build their home in this spot.

The building was structured by laying a basic grid (150cm×160cm) over the site, and using the volume of a single high-ceilinged room with a bedroom, kids' room and study partitioned off using fittings approximately half the height of the ceiling. The expanse of the entire ceiling can be felt from any room.

As the inside space is in full view from surrounding tall buildings merely with transparent toplight, the decision was firm from the beginning to install the ceiling under the toplight. Through the installation of the ceiling to make it possible to delicately feel the change and expression of the sky through it rather than directly looking up, coming up with an idea that how to configure the indoor space to sufficiently acquire the feeling of outside expression became a focus of attention. When it comes to the ceiling, it is a white acrylic vault board ceiling enabling the projection of

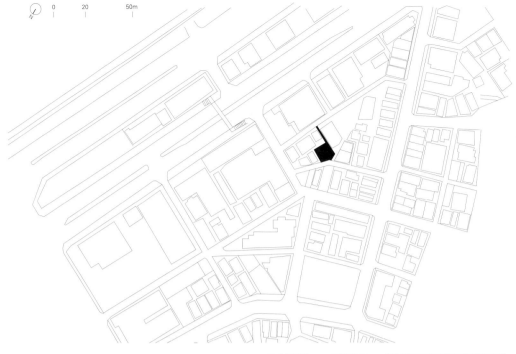

1. rooftop floor: wood deck t=20, waterproof urethane painted, drainage slope 0-60, concrete deck t=130
2. toplight: low-e glass t=4mm, air cavity t=12mm, wired glass t=6.8mm
3. interior finish: structure board t=9mm, white osmotic paint, double coating
4. ceiling: creamy white acrylic board, t=3mm
5. outer wall: galvalume silver, waterproof sheet, waterproof plasterboard t=12.5mm, structure board t=12mm, insulation glasswool t=105mm
6. floor: dustproof paint, mortar troweled finish t=100mm, insulation styrofoam t=60mm
7. crest table: galvalume silver

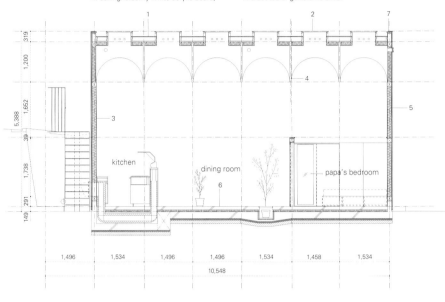

A-A' 剖面图 section A-A'

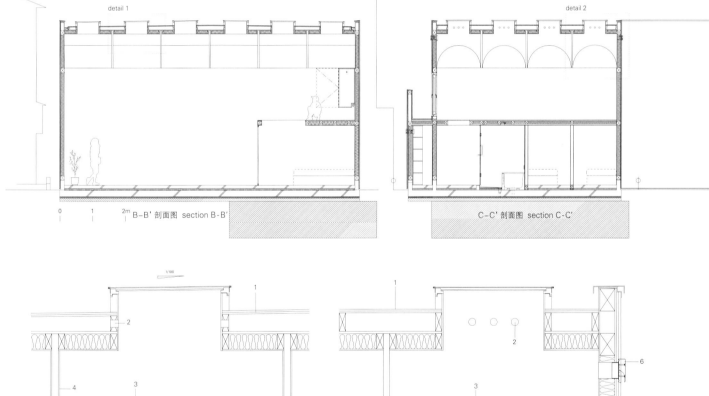

B-B' 剖面图 section B-B'   C-C' 剖面图 section C-C'

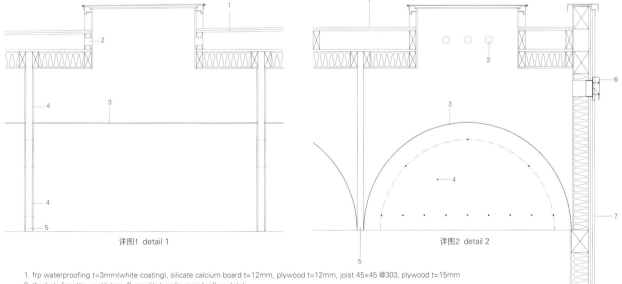

详图1 detail 1   详图2 detail 2

1. frp waterproofing t=3mm(white coating), silicate calcium board t=12mm, plywood t=12mm, joist 45×45 @303, plywood t=15mm
2. the hole for attic ventilation 3. acrylite(acrylic resin/milky white)
4. batten for acrylite 5. LVL t=45mm
6. vent cap 7. galvanised steel sheet 8. larch plywood t=9mm, penetration based paint(white)

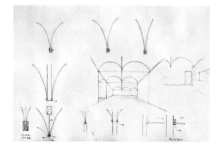

the subdued expression of light and sky coming through glass top-light on the ceiling surface. In addition to the mortar floor and larch plywood wall, the concrete white acrylic surface of high visibility is deprived of its concrete existence by its detailed treatment as if there is no brick mold. In doing so, the surface loses its concrete existence and acquires its abstractness as if only the light appears at the upside of the inside space.

There is a window (90cm×70cm) in each room. By this window, the wind blows through the whole house. The person can spend it comfortably.

There is an air space between the acrylic surface and the roof, and forced air is used to eject air heated by the sun in summer out of the building, while movement of the air is stopped in winter to use the air layer as a thermal buffer to ensure the thermal environment inside is stable.

Upon entering the building, there is so much light from the sky that it is hard to believe that the site is nested in a dark valley created by buildings. This house was named "Daylight House". Daylight does not simply indicate light from the sun, but refers to the beautiful light throughout the day, for instance, the change of the sky color when morning dawns, completely bright sky, time zone of direct sunlight, shadow of the sun by cloud movement, sky color during the time zone of sunset, moonlight and so on.

It has become ordinary at this home to feel the night sky through the ceiling when putting out light and retiring for the night. The projection of the change of 24 hours on the ceiling surface enables the home to always provide an atmosphere to be able to feel the change of outside nature even while staying inside. Takeshi Hosaka Architects

项目名称：Daylight House
地点：Yokohama, Japan
建筑师：Takeshi Hosaka
项目团队：Takeshi Hosaka, Megumi Hosaka
结构工程师：Hirofumi Ohno
甲方：Keigo Nishimoto
用地面积：114.92m² 建筑面积：73.6m²
建筑规模：two stories
高度：5.388m
结构：wooden
设计时间：2010.2
竣工时间：2011.3
摄影师：©Koji Fujii/Nacasa&Partners Inc.(courtesy of the architect)

一层 first floor

1 入口　1. entrance
2 厨房　2. kitchen
3 餐厅　3. dining room
4 起居室　4. living room
5 卧室　5. bedroom
6 卫生间　6. toilet
7 衣橱　7. closet
8 书房　8. study room
9 浴室　9. bathroom

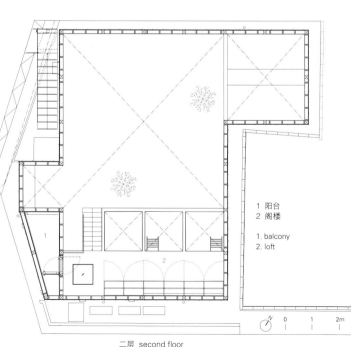

二层 second floor

1 阳台　1. balcony
2 阁楼　2. loft

内在丰富性建筑 Introvert Potential

Mecenat美术馆
Naf Architect & Design

该美术馆由Kanji Kato先生个人创立和所有,他对日本绘画风格十分感兴趣,其筹划该美术馆的目的在于,希望将他的日本绘画老师Kakudo Goami先生的作品展示给更多人欣赏。

该美术馆坐落于恬静安宁的乡村中,夜晚这里没有路灯,就会变得一片漆黑。该建筑在天色变暗的时候会自动散发出柔和的光,并照亮周围的景色。

Goami先生的作品擅于运用不同的光线产生震撼的效果,因此,建筑师决定将该建筑的设计重点放在自然采光方面;光线从天窗中柔和地散射进来,穿过白色柱状物,将光线凝聚到一层;从墙体裂缝透出来的柔光映射到室外的绿地上,不同图案的光线透过432块嵌在混凝土墙体上的平板玻璃透射出来。如此一来,该建筑内就充满了多种多样的光线。

作为一个展览空间,该建筑不需要诸如柱子和横梁这样的结构,但需要大面积的墙体。该美术馆的重要元素是将自然光线和风相结合,因此该建筑的角落切割有裂缝,但这些裂缝的大小又不至于影响到展览。这种设计形式很可能导致建筑结构非常脆弱,为此建筑师们研究出了一种合理的形状,它能保证建筑结构的稳定性,就像折纸一样,经过多次折叠,构成多个角落,并通过计算机模型和三维结构分析验证。

该建筑设计旨在将各种自然环境因素融入到展览区域中。在展览空间中最大程度地减少使用人造灯光。参观者基本都是在自然光线下欣赏这些展览作品,天气和季节的变化都会使参观者对作品产生不同的印象。设计者有意将展览空间设计得易受自然环境的影响。

建筑师希望通过这座建筑,让参观者全身心地感受到,Kakudo Goami先生创作的那种基于日本东方神韵的绘画作品的深奥而温和的氛围。

Mecenat Art Museum

This museum is funded and owned personally by Mr. Kanji Kato who is particularly interested in Japanese-style painting. He had been planning a museum with the hope of exhibiting the works of his Japanese-style painting teacher Mr. Kakudo Goami to a greater public.

The site is found in a calm and peaceful rural scene, where it becomes completely dark at night with no streetlights. This building is automatically lit up when it becomes dark, softly casting light to the surrounding.

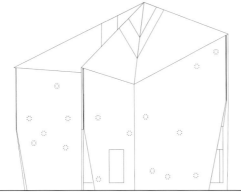
东立面 east elevation

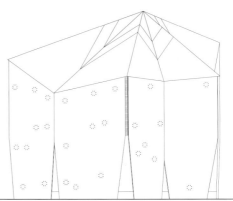
南立面 south elevation

The works of Goami give striking image of various lights, thus, the architect decided to design the building focusing on natural light; soft diffused light from top light, direct light pouring from top light through white cylinder for condensation of light spreading on the first floor, soft light from slits on the walls reflecting on the exterior green, fragments of graphical lights pouring through 432 plate glasses on the concrete walls. With this, the space is filled with various kinds of lights.

As an exhibition space, there was no need for structure such as columns and beams but as large wall as possible. The important factor of the museum was to incorporate natural light and wind, so corners are sliced with slits to the extent which would not interfere the exhibit. This idea would have left the building structurally fragile, so architects studied a rational shape, which was structurally stable like folding one sheet of paper, origami, many time to make several corners, with models and three-dimensional structure analysis by computer.

The design was focused on taking in as much factors from natural environment as possible in the exhibition space. Artificial lights in the exhibition space are limited to the minimum. The works are basically viewed with natural light which changes throughout the year, giving different impression by the weather of the day and time of the year. The exhibition space is intentionally designed to be susceptible to the natural environment.

The architect would like the visitors to feel, with their entire bodies, through this building, atmosphere of profundity and gentleness created by the paintings of Kakudo Goami, which is based on noble spirit of the Orient in Japan.

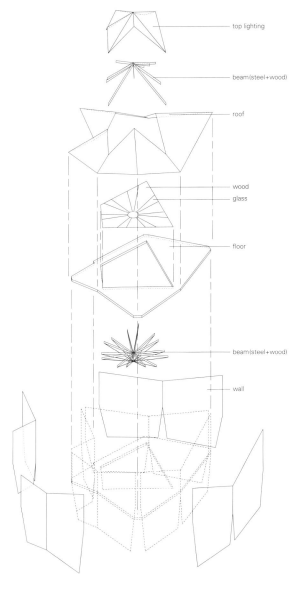

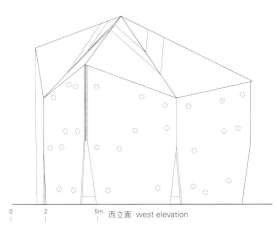

西立面 west elevation

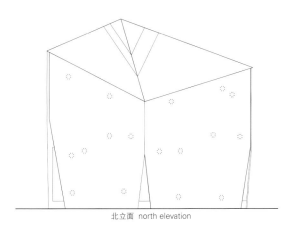

北立面 north elevation

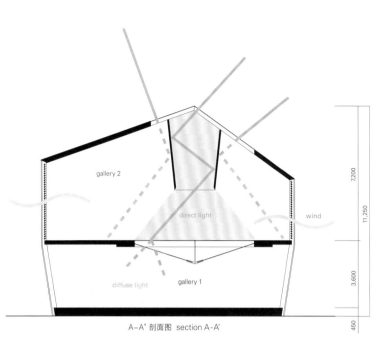

A-A' 剖面图 section A-A'

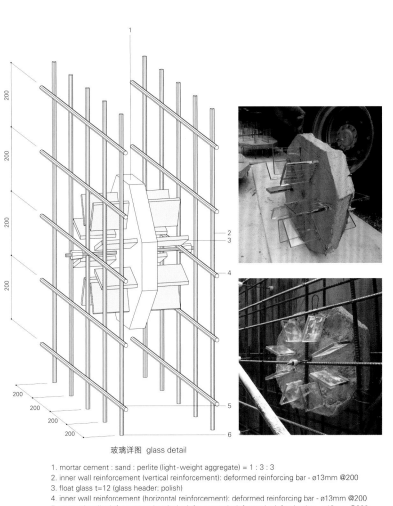

玻璃详图 glass detail

1. mortar cement : sand : perlite (light-weight aggregate) = 1 : 3 : 3
2. inner wall reinforcement (vertical reinforcement): deformed reinforcing bar - ø13mm @200
3. float glass t=12 (glass header: polish)
4. inner wall reinforcement (horizontal reinforcement): deformed reinforcing bar - ø13mm @200
5. external wall reinforcement (vertical reinforcement): deformed reinforcing bar - ø13mm @200
6. external wall reinforcement (horizontal reinforcement): deformed reinforcing bar - ø13mm @200

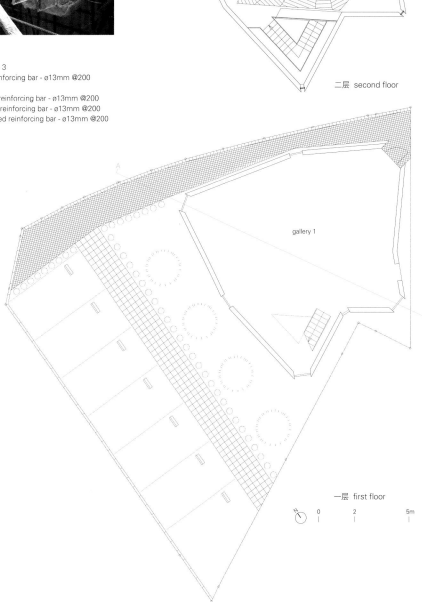

屋顶 roof

二层 second floor

一层 first floor

0 2 5m

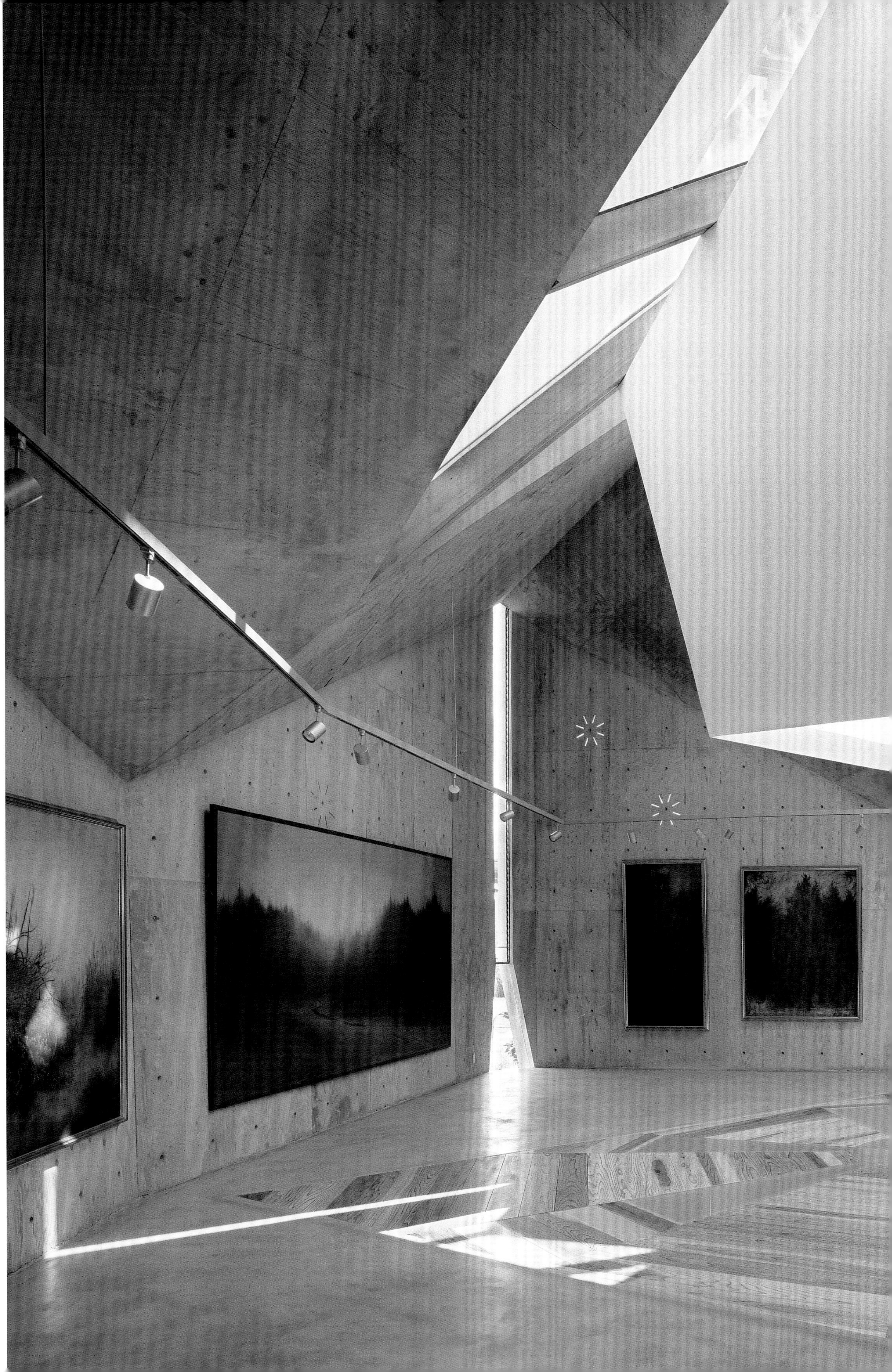

项目名称：Mecenat Art Museum
地点：Hiroshima, Japan
建筑师：Tetsuya Nakazono
工程师：Kenji Nawa
规划：Museum of Japanese-style painting
用地面积：344.57m²
建筑面积：99.55m²
总建筑面积：191.24m²
设计时间：2006.2—2009.3
施工时间：2009.4—2010.9
摄影师：©Toshiyuki Yano (courtesy of the architect)

项目名称：Mecenat Art Museum
地点：Hiroshima, Japan
建筑师：Tetsuya Nakazono
工程师：Kenji Nawa
规划：Museum of Japanese-style painting
建筑面积：99.55m²

Moliner住宅
Alberto Campo Baeza

内在丰富性建筑　Introvert Potential

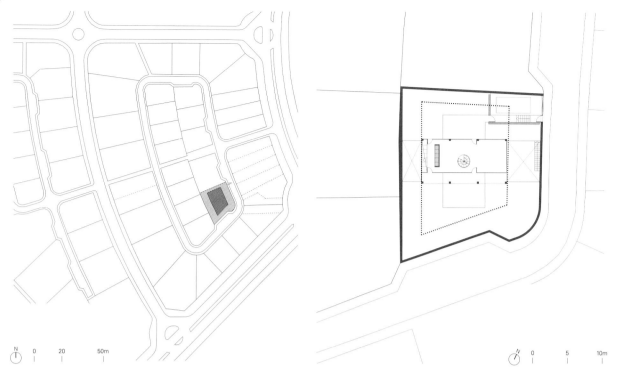

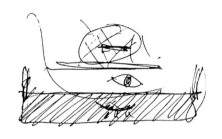

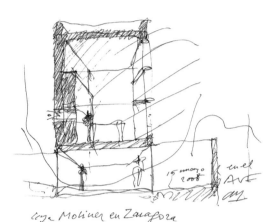

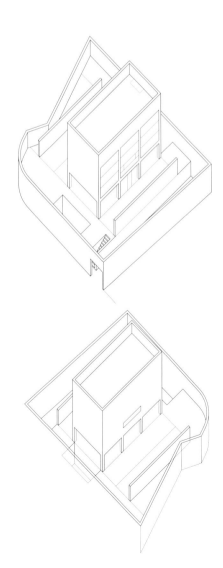

该建筑是为一位诗人而建。他可以在这里读书、写作和思考，并在此经历生老病死。

一座理想的住宅应该是让人魂牵梦绕的地方，人们很难离它而去，同时住在里面会感到很开心。

该住宅被传统房屋围绕着，建筑师们决定创造一个另类的世界——一个专属于诗人且拥有独特景观的世界。建筑师建造了白色墙体，仿佛是一个盒子，开口朝向天空，就像一座毫无装饰的超自然花园，这里有山、有水，还有郁郁葱葱的树木。

位于中心位置的是一个三层高的白色大盒子，是梦想的最高处、生活的花园，以及休息的最静谧处。

为了梦想，建筑师们在最高点设计了一层"云"。图书馆被建造成双层高。北面的光照用于读书、写作、思考和感受。南面墙体上设有通高的书架。在书桌的高度上有一扇细长的窗户朝向花园开放，阳光也可以透过它直接照射进来。这使得很小的空间看起来也很宽敞。

为了生活，花园里满是来自南面的阳光。这个空间里到处都是花园；透明的墙体和由相同石头铺成的地面，将室内与室外衔接起来。在那里，有一处与天花板和侧墙都不相接的独立结构——厨房，它位于起居室和餐厅的前面。

有一处用于睡觉，也可供安息的静谧空间位于卧室下面，犹如一个洞穴。建筑师采用了英式庭院的挖掘方式，因此人们可以在这里看到天空。

一个螺旋式楼梯将这三层连接起来，该楼梯位于房屋的几何中心处。这样当人们从中间的花园层进入屋内时，上下楼都很方便。这里还设有一个大型的餐用升降机，从而机械地将这三层连接起来。

该住宅的内外都是白色的，其内外的地面也都采用了暗淡的石灰岩。

再一次彰显了这座宛如洞穴和船舱的住宅。

这是一个梦想、生活和安息的地方，一座为诗人而建的住宅。

Moliner House

This project is to build a house for a poet, to make a house for dreaming, living and dying, a house in which to read, to write and to think.

The ideal house is the one you always want to return to, the one which is difficult to leave behind and in which you are and you feel happy.

Surrounded by other houses of conventional design, architects decided to create a special world, a world apart for a poet, with its own landscape. They raised white walls to create a box open to the sky, like a nude, metaphysical garden, with stone, water and leafy trees.

And a white box with three levels was placed in the centre. It is the highest for dreaming, the garden level for living, and the deepest level for sleeping.

For dreaming, architects created a cloud at the highest point. A library constructed with double height, with northern light for reading and writing, thinking and feeling. The southern wall is covered with shelves of books from floor to ceiling. There a little elongated window at table-level opens out to the garden through which sunlight enters directly. The dimensions and proportions of the room are such that, while small, it appears very large.

For living, the garden is full of southern light, sunlight. The space is all garden, with transparent walls and the same stone floor that brings together inside and outside. There, behind a free-standing piece that joins neither the ceiling nor the sides is the kitchen. It is in front of the living and dining room.

And the place for sleeping, perhaps dying, is in the deepest level, which is below the bedrooms, as if in a cave. It was excavated like an English-style courtyard so that one can see the sky.

The connection between the three floors is a spiral staircase situated at the geometric centre of the house. As one enters the house via the middle garden floor, access to the upper and lower floors is immediate. There is an elevator which is a large dumb waiter that mechanically links up the three floors.

The house is white inside and out. The floors both inside and out are all pale beige limestone.

Once again, the cave and the cabin.

Dreaming, living, dying. The house of the poet. Alberto Campo Baeza

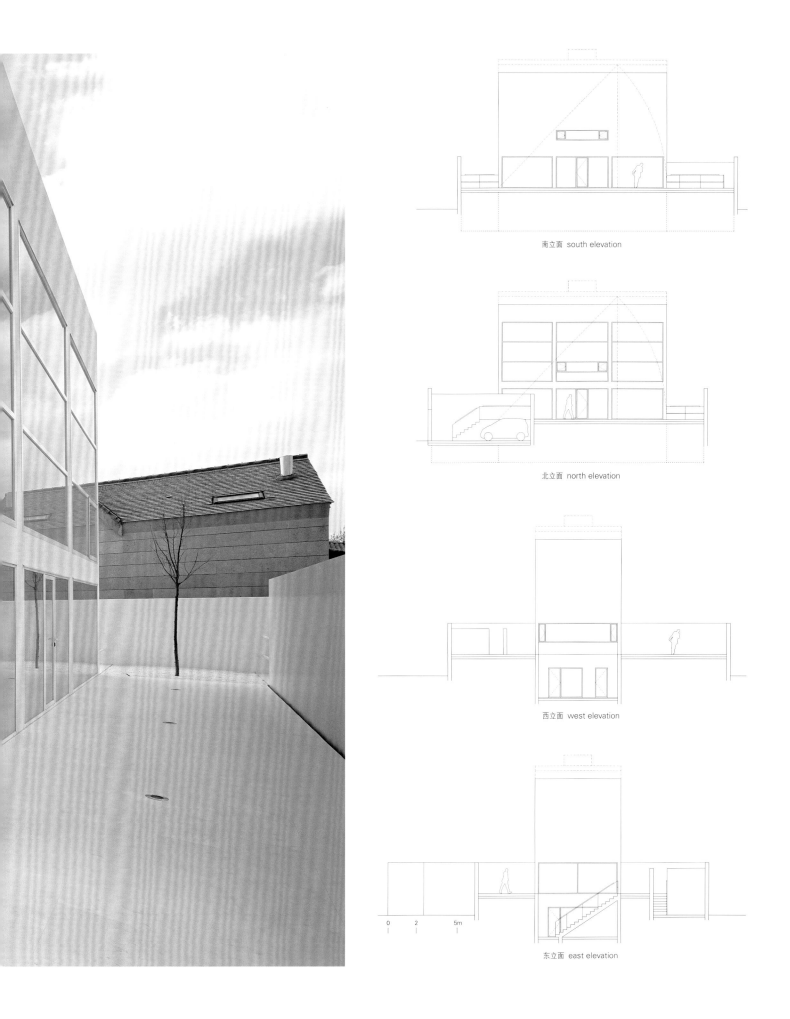

南立面 south elevation

北立面 north elevation

西立面 west elevation

东立面 east elevation

项目名称：Moliner House
地点：Avda. Ilustración, 40, Urbanización Montecanal, Zaragoza
建筑师：Alberto Campo Baeza
合作建筑师：Ignacio Aguirre López, Emilio Delgado Martos
结构工程师：María Concepción Pérez Gutiérrez
施工技术员：José Miguel Moya
设备工程师：Saneamientos Delicias - Daniel Laborda
电力工程师：Monvier, Lamp, Años Luz
承包商：Construcciones Moya Valero-Rafael Moya, Ramón Moya
开发商：Luis Moliner Lorente
用地面积：445.50m²
建筑面积：216m²
设计时间：2006
竣工时间：2008
摄影师：©Javier Callejas(courtesy of the architect)

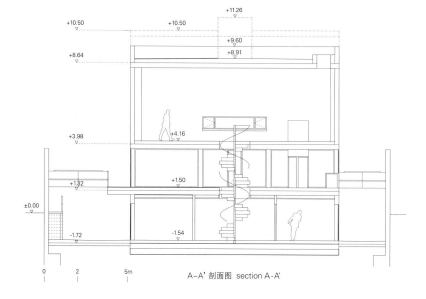

A-A' 剖面图 section A-A'

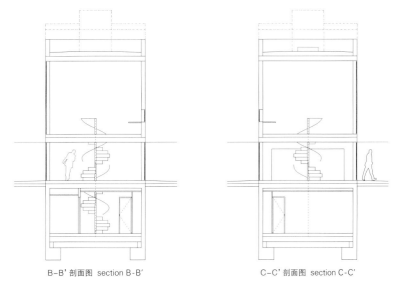

B-B' 剖面图 section B-B'　　　　C-C' 剖面图 section C-C'

1 图书馆　1. library
二层　second floor

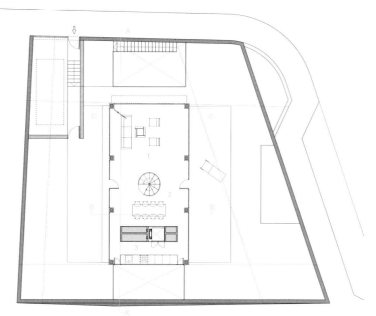

1 起居室　2 餐厅　3 厨房　1. living room　2. dining room　3. kitchen
一层　first floor

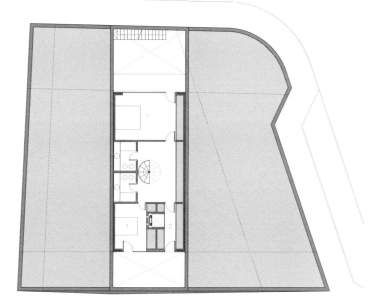

1 卧室　2 洗衣房　1. bedroom　2. laundry
地下一层　first floor below ground

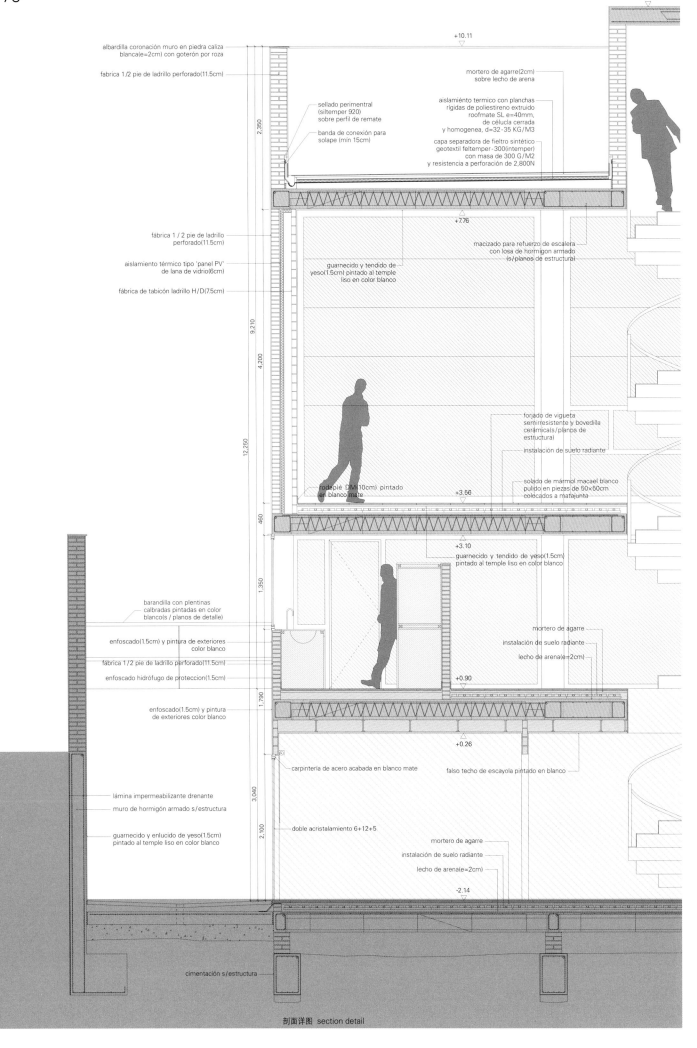

剖面详图 section detail

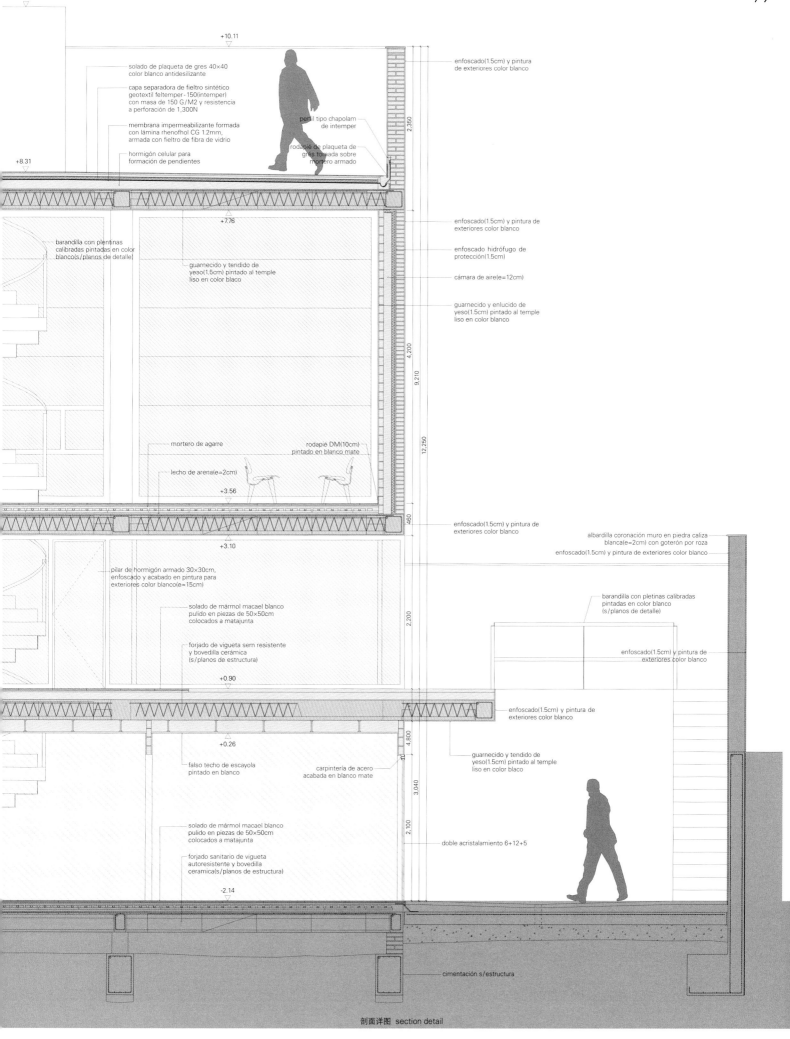

剖面详图 section detail

城市遗产

City Inherit

城市是各种各样相关元素的构造物，由社会经济力量驱动，随时间做出动态改变。社会学家克劳德·列维·斯特劳斯把城市描述成"自然与人工制品汇合而成的、最复杂的人类介入成果"（1954年）。

我们将通过对三个不同案例的回顾来论述欧洲现代规划和设计方式中的问题。所有项目都位于城市的核心——历史文化中心。现存城市环境中的项目成为沟通城市规划与建筑的桥梁。

这里我们尤其注意到扩建与城市之间的关系。建筑物是否真的改变了该地区的本质（地域特色），每个项目又是如何为更适于居住的环境和更积极的城市形象（意象）做出贡献的？

人们引入城市重建和"城市针灸"的现代规划实践来评估建筑物背后的过程和意义。成功的城市设计运作通过城市环境的可识别性高低进行考量。

Cities are a construction of various interrelating elements, driven by socio-economic forces, transforming dynamically over time. Sociologist Claude Levi-Strauss described the city as *"the most complex of human interventions…at the confluence of nature and artifact"* (1954).

Three different cases will be reviewed to address issues of current planning and design approaches in Europe. All projects are located in the heart of the city, the historical center. Projects within the existing city context have to bridge the gap between urban planning and architecture.

Special attention is given to the relation between the addition and the city. Does the building change the nature of the place (Genius Loci)? How does each project contribute to a more liveable environment and a more positive image of the city (Imago Urbis)?

The current planning practices of urban regeneration and urban acupuncture are introduced to evaluate the process and the meaning behind the architectural reference. The successful exercise of urban design is measured in regard of the legibility theory of an urban environment.

贝尔法斯特都市艺术中心/Hackett Hall McKnight
隆德大教堂广场建筑/Carmen Izquierdo
圣安东慈善学校修复工程/Gonzalo Moure Architect
城市遗产的填充/Michele Stramezzi + Maria Pedal

MAC, Belfast/Hackett Hall McKnight
Lund Cathedral Forum/Carmen Izquierdo
San Antón Charity School Restoration/Gonzalo Moure Architect
Heritage Infills/Michele Stramezzi + Maria Pedal

城市设计：城市遗产的填充

城市复兴

城市复兴不仅仅是简单地在某一项目地址上建造建筑，而是一种"综合体，是视觉与行动的结合体，能够解决城市问题和寻求为被改变地区带来经济、物质、社会和环境条件的持久改善。"——彼得·罗伯茨与休·塞克斯，《城市复兴》，指南，伦敦：SAGE出版社，2000年，第17页

几十年来，欧洲都在进行城市复兴的实践活动，其目的在于解决城市结构问题，尤其是空间问题。直到21世纪初，城市复兴才被发展成为一种综合的多层次策略。其规模和影响从地方性的社区建设方式一直延伸到区域性，甚至是全国性的策略。

复兴和增强城市生活的活力以及公共区域活跃性的创新性方法就是理论术语上的"城市针灸"。与城市复兴不同，"城市针灸"的范围仅限于较小规模，以具有催化效果和直接环境的小规模、策略性介入为目标。这些介入通常通过比普通城市复兴过程更少的时间来实现；整体视觉效果也不是那么复杂。然而两种实践的目的都是要超越建筑的层次，为被改建的地点实现最大化利益。

"当城市介入不得不在统一的城市组织中实施的时候，它就会受到公共空间范围的限制。接着'城市针灸'的概念就可以被改成'公共空间针灸'，我们可以将其定义为城市公共空间复兴和基于独立的、协调的小规模介入的城市生活复兴的一种有效城市策略。"——Salomon Frausto，贝拉罕建筑研究所，荷兰代尔夫特

"城市针灸"工程不是整体视觉效果的一部分，也不属于系列战略行动，但可以被比作城市复兴的关键工程。

本文将检验三个不同的现代城市规划与设计案例，这些项目属于城市复兴或"城市针灸"范畴，它们之间的关系是很明显的；同时它们的建造地点和使用目的又很特别。

文中所分析的项目所在的城市分别位于欧洲大陆的三个角落——从瑞典和爱尔兰到西班牙。

功能

城市中心从整体上代表着某一社区及其居民，在城市生活中起

Urban How: Heritage Infills

Urban Regeneration

Urban regeneration is more than the plain application of architecture to a site, it is a *"comprehensive and integrated vision and action which leads to the resolution of urban problems and which seeks to bring lasting improvement in the economic, physical, social and environmental condition of an area that has been subject to change."*
– Peter Roberts and Hugh Sykes, *Urban Regeneration*, A Handbook, London: SAGE Publications, 2000, p.17

For decades the practice of urban regeneration has been carried out in Europe aiming to tackle urban issues mainly with spatial solutions. Only since the beginning of the XXI century urban regeneration has developed into an integrated and multi-layered strategy. The scale and impact of urban regeneration programs can stretch from a local neighbourhood based approaches up to regional or national strategies.

An innovative approach to regenerate and enhance vital urban life and active public places can be theorized with the term "urban acupuncture". Different from urban regeneration, the extent of urban acupuncture is limited to the lower scale, targeting strategic small-scale intervention with a catalytic effect or the immediate environment. These interventions are usually realizable in a shorter amount of time than the average urban regeneration process; the overall vision is less complex. Nevertheless both practices aim to achieve the maximum benefit for the location due to transform beyond the level of architecture.

"When the urban interventions have to be implemented in the consolidated urban tissue, they frequently have to be limited only to the ambit of the public space. Then the notion "urban acupuncture" can be transformed into "public space acupuncture", which can be defined as an effective urban strategy of regeneration of the public space of the city and its urban life based on independent, but coordinated small interventions." – Salomon Frausto, Berlage Institute, Delft, NL

Urban acupuncture projects can be compared to key projects in urban regeneration without being part of an overall vision or a set of strategic actions.

The article will examine three contemporary cases of urban planning and design; the relation of this works within the boundaries of urban regeneration or urban acupuncture is obvious; their location and destination of use are peculiar.

The analyzed projects are located in cities on three corners of the European continent, from Sweden and Ireland to Spain.

Function

Representing the community and its inhabitants as a whole, the city center plays a specific role in the urban life. The center is

到特殊作用。城市中心的公共领域能够作为活跃的空间来体验：它不但为使用者提供其所需，同时也激励和激发其相互作用，支持娱乐活动，它让人记忆深刻又与城市本身的其他地方有所不同。

多功能性是对一个成功社区的相关考量方式：一个具有吸引力和混合功能的项目是公共社会生活具有活力的基础。

环境—形态学

从大体上说，欧洲城市中心的形态学定义过程早在几世纪前就已完成，其发展的层次可追溯到中世纪，甚至古罗马时期。城市街区的形态由之前存在的城镇空间所定义，是城市规划与设计的一个基础部分。

理论

基于对城市形式的定义，我们通过不同的方式观察城市空间及其特点，它们可以被粗略地划分为具有吸引力和不具有吸引力的。

为什么我们体验到一些地方与其他地方不同？可识别性概念就是其原因所在。可识别性理论可以追溯到美国城市专家凯文·林奇的研究。在1959年出版的《城市形象》一书中，这位麻省理工学院的教授将可识别性或形象性定义为"一种特性，它使一个有形物体能够为任意观察者带来一种唤醒强烈形象感的高度可能性……在这一特殊意义上，一个具有高度形象性的城市应该完整、清晰、别具一格，能够让眼睛和耳朵更集中注意力，并更有效地参与。"

在城市构成中，一个清晰但不显著的建筑对这一地方的可识别性是有益的。一个极具吸引力的城市在很大程度上依赖于其最重要的城市元素——道路。其他的要素如果能够合理布局，并带有意义、可辨别性和清晰性，也能够为城市环境的高度可识别性做出贡献。

城市街区是城市构成的最小实体，它被塑造城市结构的街道所包围。

我们所检验的项目无一不在城市街区范围内填充、围合或完善着城市结构中的一个空间：每一个单独的建筑物都不仅仅是建筑而是城市的一部分，与公共区域相关并与之呼应。

这种发展及其实施所面临的一项巨大挑战，就是如何将建筑介

the place (Locus) where the public domain can be experienced as a vital space: it offers more than just necessities to its users. It inspires and stimulates interaction, supports recreation, and is memorable and distinctive from other places within the city itself. A relevant measure for a successful neighborhood is its multi-functionality: an attractive mixuse program is the basis for a vital public social life.

Context-Morphology

In general the process of morphological definition of the European city center has been completed centuries ago, often providing layers of expansion dating back to medieval or even Roman periods. Morphologically defined by the pre-existing townscape, the urban block is the elementary part of city planning and design.

Theory

Depending on how the urban form is defined, we perceive the urban space and its character in different ways, roughly divided between attractive and unattractive.

Why we experience some places different than others can be explained within the concept of legibility. The theory of legibility goes back to the studies of American urbanist Kevin Lynch. In his book *The image of the City*, from 1959, the MIT Professor defines legibility or imageability as *"that quality in a physical object that gives it a high probability of evoking a strong image in any given observer… A highly imageable city in this peculiar sense would seem well formed, distinct, and remarkable, it would invite the eye and the ear to greater attention and participation."*

A clear but not obvious structure of the urban composition is beneficial for the place legibility. An attractive city is greatly dependent on the most predominant city component, the paths. The other elements, if well placed and with strong attributes as meaning, distinctiveness and clearness are contributors to a highly legible urban environment.

The area of the urban block is the smallest entity of urban composition, surrounded by streets that shape the urban fabric.

Each of the examined projects either fills, closes or completes a void in the urban fabric on the scale of the urban block: every single building is more than just architecture but is a piece of city, relating and responding to the public realm.

A major challenge of such a development and its implementation is how the built interventions relate to the local identity and how they can add qualities to the urban environment. The most desired outcome in any case of urban intervention is to result in a

隆德大教堂广场建筑通过与周围城市景观的规模和线条相适应,来融入城市结构。
Lund Cathedral Forum integrates itself in the urban fabric by adapting to the scale and lines of the surrounding cityscape.

入与地方特性相结合,又如何在城市环境中加入新的特质。在任何城市介入的案例中,人们最渴望的成果是产生地方协同效果,在那里单独元素产出的总量比先前投入的更多。这种设计支持给定的元素相互作用,以改善先前的环境,使其向更强烈的地方感转变。每个项目都要填补城市设计与建筑之间的缺口。关注的重点是与城市的关系;建筑和城市化是和谐统一的关系吗?扩充建筑会改变一个地方的本质(地域特色)吗?每个项目又是如何为创建一个更生动的环境和更积极的城市形象(意象)做出贡献的?

Carmen Izquierdo设计的隆德大教堂广场建筑

瑞典南部的隆德镇拥有8万人口,是该国最古老的城市之一;然而大部分的城市中心建筑都可以追溯到19世纪,同时也混杂着一些中世纪的遗迹,隆德大教堂就是其中之一。

隆德城市中心最新的扩建项目就是由建筑师Carmen Izquierdo设计的隆德大教堂广场建筑,于2010年到2011年施工完成。它位于隆德大教堂对面,处在该城市最具历史意义的地点之一。

新建的广场建筑空间由那些原有的历史建筑物构成,形成了一个重新修建、连成一体的城市街区。

项目的线条和外形混合了那些现存城市环境,保留了对周围公共区域可识别性的改进。

其成果在于中间一半古老、一半现代的空间,其中带有玻璃外立面和最小型白色室内空间的新建筑与现存建筑相映成趣。

教堂广场建筑与周围街道之间的关系将从前未知的强度引入了该区域;建筑在各个侧面都运用了不同的方法:一面是由封闭的围墙所定义的三角形广场;立面内凹处将人们引至主要入口,人工制作的顶部开放的屋顶仅对街道的外观设计起作用;最终扩建的入口在小庭院之中,面向教堂。

广场与教堂的现实关系暗含在一个特殊的建筑姿态中——"视觉框架"这一出人意料的元素。这一元素减少了两座相对建筑之间在视觉和空间上的距离感。礼堂由带有雕刻的空间装饰,而巨大的方形窗户可远眺大教堂。从建筑外部可以看出,这是最强有力的构成选

local synergy, where the sum of the single elements is a larger outcome than the original input. The design should support an interrelation amongst the given components to generate an improvement of the previous situation towards a stronger sense of place. Each project has to bridge the gap between urban planning and architecture. Center of attention is the relation to the city; do architecture and urbanism relate in a uniting fashion? Does the built addition change the nature of the place (Genius Loci)? How does each project contribute to a more liveable environment and a more positive image of the city (Imago Urbis)?

Lund Cathedral Forum by Carmen Izquierdo

The eighty thousands inhabitant's town Lund in the south of Sweden is one of the oldest cities in the country; yet most of the city center architecture dates back to the nineteenth century mixed with several medieval remains. One of these relics is the Lutheran cathedral.

The latest addition to lund city center is the Lund cathedral forum by architect Carmen Izquierdo, realized over the period from the year 2010 to 2011. It is located just opposite the church in one of the most meaningful historical places of the city.

The new forum composes its volumes with those of pre-existing historical buildings, resulting in a restored and coherent urban block.

The lines and dimension of the projects blend in with those of the existing urban context, preserving and improving the legibility of the surrounding public realm.

The result are the inbetween spaces, half old half contemporary, where the new-built architecture faces the existing with glass facades and minimal white interiors.

The relation of the cathedral forum with the surrounding streets introduces a before unknown intensity to the location; on all sides the building offers a different approach: a triangular square on one side defined by closed walls; a recess in facade as an invitation towards the main entrance, with the artifact of a top open roof with the mere function of the street profile design; a final additional entrance in a little court facing the cathedral.

The physical relation with the cathedral is underlined with one specific architectural gesture, the surprising element of the "Visual Frame". The element reduces the spatial and visual distance between the two opposed buildings. The auditorium finishes with a sculptural volume with a large square shaped window overlooking the cathedral. Judging from the building's exterior this is the

都市艺术中心被看作公共领域的延伸。
MAC is considered as an extension of the public realm.

择,高耸的空间在结构上占据了大教堂的形象,将教堂收纳在广场建筑之中。

新建广场建筑为建筑的可识别性、向导性和外形特征做出了贡献,同时将城市环境的线条和外形与雕刻的建筑语言相搭配,将这个历史小镇引入了21世纪。

贝尔法斯特都市艺术中心,哈克特·豪尔·麦克奈特建筑师事务所

贝尔法斯特新建的博物馆占据了城市结构中与众不同的一个地点,以其紧凑的体量为特征。该项目位于北爱尔兰首都的中心和具有历史价值的新兴大教堂区中部。2007年,爱尔兰哈克特·豪尔·麦克奈特建筑师事务所赢得了英国皇家建筑师协会国际设计大赛,开始设计建造MAC——贝尔法斯特都市艺术中心的简称。从程序上,这项工程包括展示空间、艺术品展览馆和支撑设施。

这座建筑在重建该区域和将整个城市从原先的主要社会和政治问题及冲突中复原的策略中起到了关键作用。2012年初,该市就开展了全新的美德运动。

这座复合建筑被插入到圣安妮英国大教堂后面的狭窄街道之间。贝尔法斯特都市艺术中心顶部采光的多层休息大厅将城市的公共空间扩展到了建筑内部。

最终建成的建筑是对先前该区域内维多利亚遗迹的回忆,而结构细部的层次是现代建筑的最小语言。

建筑体量限制在两座现存建筑之间,填补了城市结构的空白。

设计的目的在于将狭窄的周边街道融入城市环境,并将使用者带入新建的圣安妮广场,都市艺术中心的主入口正设在那里。这一城市建筑为城市公共区域开辟了新的领地,同时将带领居民和观光者参与到城市规划的文化扩建中。

这一建筑的主要特色之一在于比周围建筑都要高的塔楼。塔楼使都市艺术中心在周边的建筑中显而易见,成为贝尔法斯特城市结构的一处地标。

塔楼不但从远处可见,同时也为都市艺术中心的观光者提供了一幅崭新的城镇空间透视图。

strongest composition choice, a raised volume capturing the image of the cathedral in a frame, bringing the church inside of the forum itself.

The new-built forum contributes to legibility, guides, frames, while matching lines and dimension of urban context with modern sculptural architectural language, and introducing 21st century to the historical town.

MAC, Belfast by Hackett Hall McKnight

Belfast's new Museum occupies a rather unusual spot within the urban fabric, characterized by its compressed parameters. The project is situated in the center of the Northern Ireland capital, in the middle of the historical valuable and emerging cathedral quarter.

In 2007 the Irish architecture office Hackett Hall McKnight won the RIBA International Design Contest to design and build the MAC, short for Metropolitan Arts Center, in Belfast. Programmatically the work involves performance spaces, art galleries and supporting facilities.

The development plays a key role in the strategy to regenerate the district as well as in the rehabilitation of the entire city from its previous predominant social and political issues and clashes. Early in 2012 the city's brand new cultural virtue was opened.

The multipart ensemble is inserted in-between narrow streets in the back of St Anne's Anglican cathedral. The top lighted multistory foyer of the MAC Belfast expands the public space of the city into the building's interior.

The materialization is a reminiscence of the preceding Victorian heritage of the district, while the level of architectonical details refers to a minimal modern architectural language.

The built volumes are constrained between two existing buildings filling a gap in the urban fabric.

The design aims to contextualize the narrow surrounding streets and guides the users to the freshly established St Anne's square, where the main entrance of the MAC is placed. This urban composition reclaims new urban grounds for the city's public realm and at the same time invites inhabitants and visitors to engage the cultural addition to the urban program.

One of the main architectural features of the composition is the tower whose height reaches over the surroundings. The tower marks the address of the Metropolitan Arts Center beyond the boundaries of the neighborhood, contributing a landmark to the urban structure of Belfast.

圣安东慈善学校,绿色的开放式公共入口成为建筑综合体的核心。
San Antón charity school, green open public access is set as core of the complex.

由Gonzalo Moure建筑师事务所设计的圣安东慈善学校修复工程

一场大火摧毁了圣安东慈善学校原本的城市街区,该场址中新的多功能综合楼由位于西班牙首都马德里中心的、当地的Gonzalo Moure建筑师事务所设计完成。

2005年,众多的股东带着想要修复城市这一部分的共同愿景聚集到一起。核心意见就是要插入一个社会性项目,利用其足够大的影响力来刺激周围区域的复兴。项目的混合功能从日托所一直到带有游泳池的运动设施,此外还包括马德里建筑师学院(COAM)和COMA文化协会。中央公园为(周边)各种不同活动提供空间。

这个已建成的项目可谓是一个考虑周密、与周边和谐统一的介入体。作为一个填充物,如果没有原有建筑的附加表层,该建筑与城市街区并不相关。被破坏的建筑的原有外立面得到保留,同时建筑内部又恢复了生机。

玻璃和钢塑造了新建学校建筑的外形,赋予建筑以现代感,同时又具有"不合时宜"的外观和感觉。

结论

这三个项目都承担了一座公共建筑对城市及其居民所承担的社会责任。

定义建筑外观的物理参数成为连结城市规划与建筑之间的桥梁。所有的扩建物都与先前存在的城市结构成为不可分割的整体,同时提高了区域的可识别性。每个案例的城市设计都小心翼翼地考虑到与城市融为一体的历史建筑的线条与外观。

从另一方面说,建筑的目的并不是要混合或者模仿传统的地方风格。建筑语言是当代的、现代的,大部分时候都非常简洁且不具有地方性参考。文中所述项目在地理上无法定义彼此间的不同。

当今的欧洲教育建筑似乎不再通过区域性或国家的传统特色进行识别。

The tower cannot only be seen from a distance but also allows at the same time visitors of the MAC a new perspective over the townscape.

Restoration San Antón Charity School by Gonzalo Moure Architect

A fire destroyed the original urban block of San Antón charity school, and the site of the new mixuse complex designed by local architectural office Estudio Gonzalo Moure, is in the heart of the Spanish capital Madrid.

In 2005 multiple stakeholders came together with the shared vision to restore this part of the city. The central idea was to inject a social program with impact large enough to stimulate a revitalization process for the surrounding district. The mixuse program stretches from a day care to sport facilities with swimming pool, furthermore it includes the headquarter of the Colegio Oficial de Arquitectos de Madrid(COAM) and the COAM Cultural Foundation. The central garden offers room for different (neighbourhood) events.

The realized project is a very careful and coherent intervention. Working mainly as an infill, the building doesn't relate to the exterior of the urban block if not for an added layer to the original structure. The original facade of the destroyed building is maintained and the inner part of the block is re-activated.

Glass and steel shape the volume of the new school building, giving the architecture a sense of contemporary as well as an "out of time" appearance and feeling.

Conclusion

Each of the three developments shares the social responsibility that a public building has towards the city and its users.

The physical parameters that define the built volume aim to bridge the gap between urbanism and architecture. All additions form an integrated part with the pre-existing urban structure joined by an improved effect on the area's legibility. The urban design of every example refers carefully to the lines and dimension of the historical context shaping a coherent part of the city.

The architecture on the other hand is not intending to blend in or imitate the traditional local style. The architectural language is contemporary, modern and most of the time minimal without any vernacular reference. The reviewed projects are non-definable from each other's difference in geography.

Educated European architecture of today seems not to identify with regional or national customary characteristics.

Michele Stramezzi + Maria Pedal

贝尔法斯特都市艺术中心
Hackett Hall McKnight

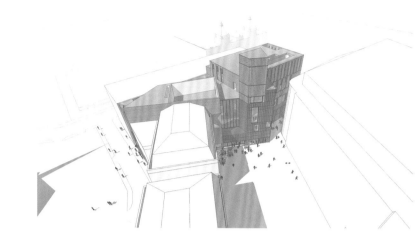

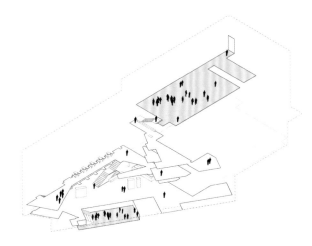

在竞标阶段,建筑师们借鉴了Ciaran Carson的诗歌"Turn Again"。它讲述了一个古老的城市——贝尔法斯特。他们想使该建筑与老贝尔法斯特融为一体。该建筑不同于大多数当代建筑,建筑师们想要使它给人一种很旧的感觉,同时可能让人感到很新奇。就这一点而言,它的周边环境显得并不协调。眼下新建的圣安妮广场,原本是要显示它的古老,却并没能显示出它的历史内涵。所以,建筑师转向了那些维多利亚时代的砖城(尤其是仓库)所具有的更真实的特征。也就是说,他们将这一新广场看作该城市新增的一处具有积极作用的公共场地。建筑师通过设计一座塔楼来了解这一空间并与它进行融合,这座塔楼像钟楼一样,成了这个广场的地标性建筑。在他们看来,这座塔楼带给人们一种先于场地上一切建筑的感觉,它的坚固性和结构重量展现了岁月感和永久性。

除了塔楼以外,这个项目还有两座砖石建筑,它们采用场地上的混凝土建造而成,并涵盖了主要的空间。它们用砖石覆层,且通过对建筑语言的微小调整表达了各自的特征。其中一种是通过重复的立面处理手法形成的规则长方体形式,展现用小型住宅单元覆盖大空间,另外一种是大型独立开口的较不规则的形式,提供了更大、更宽阔的空间。

在两座砖石建筑之间是高大的门厅。它唤起了人们对邻近狭窄街区的记忆。那是一个很拥挤的空间,特色是顶部照明,内部是砖和混凝土立面。由于场地的限制,建筑被迫采用堆叠式的结构,走廊建在剧院的上面,像"高地"一样,人们走上围绕门厅而建的楼梯可以到达那里。

为了清晰地表达建筑理念,其中一个混凝土盒式结构的砖石覆层已经被剥掉了,露出了围绕门厅的五层楼高的混凝土墙。建筑师们在项目的最早期就提议将这面墙装饰成模板纹表面的混凝土墙。随着项目的进行,他们又重新考虑了这种对墙的处理方式。他们意识到,模板纹这一提议并不是出于建筑要求,而是出于一种创造纹理并引起人们兴趣的美学直觉,来使场地的大型表面变得生动,使其看起来有

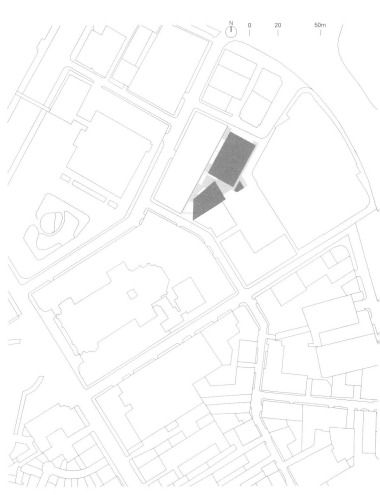

破碎感。这种意识引导建筑师们考虑怎样将这种美学直觉延伸到一些建筑表面的图案或纹路上。对砖质立面的研究,促进了这一手法在类似地势上建造的石墙或者混凝土墙的发展。通过将光滑的混凝土框架用到带有大量纹理的模板纹混凝土中,有意将图案引入混凝土墙被描述成一种抽象绘画。塔的外部表面也使用了这种"画出来的"或者"有纹理的"立面手法,这样,塔楼的玄武岩表面就展现出一种起伏不平的条状图案,凹进广场。

都市艺术中心作为城市中心的新文化大楼是与众不同的,因为很多室内的空间可以作为公共区域。正是这个原因,建筑师希望都市艺术中心可以带给人们他们所预期的体验,成为贝尔法斯特真正的公共场所。

MAC, Belfast

During the competition stages of the project we referred to the poem "Turn Again" by Ciaran Carson. It speaks of the old city – an older Belfast with which we wanted the building to feel connected. Unlike much contemporary architecture we want to make buildings feel old as much as they might feel new. In respect of this the context was difficult. The immediate presence of the new St.Anne's Square, whilst designed to appear aged, did not have a "depth" to its apparent history. So, instead we looked to the more authentic characteristics of the Victorian brick city and, particularly, its warehouses. That is, we viewed the presence of the new square as a positive addition of a new public space for the city. Our way of understanding and engaging with this space was through the design of a tower which addresses the square – a campanile-like gesture. The tower, in our reading, has a sense of predating everything else on the site, its solidity and constructional weight conferring a sense of age and permanence to the space.

In addition to the presence of the tower the project is defined by two brick blocks, constructed from in-situ concrete, which contain the main spaces. Clad in brick, the expression of each block is distinctive due to minor adjustments to the architectural language. One is a regular cuboid form expressed with a repeated elevational treatment that reflects the wrapping of large spaces

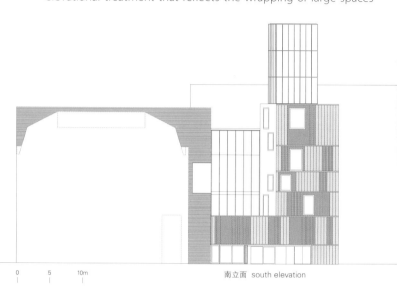

南立面 south elevation

东南立面 southeast elevation

东立面 east elevation　　西南立面 southwest elevation

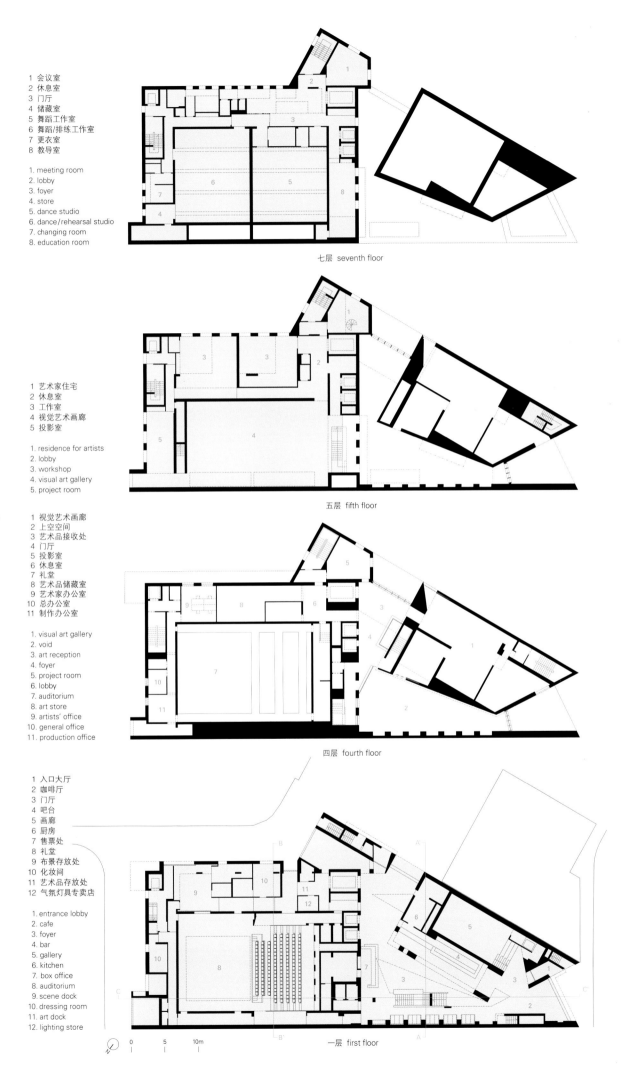

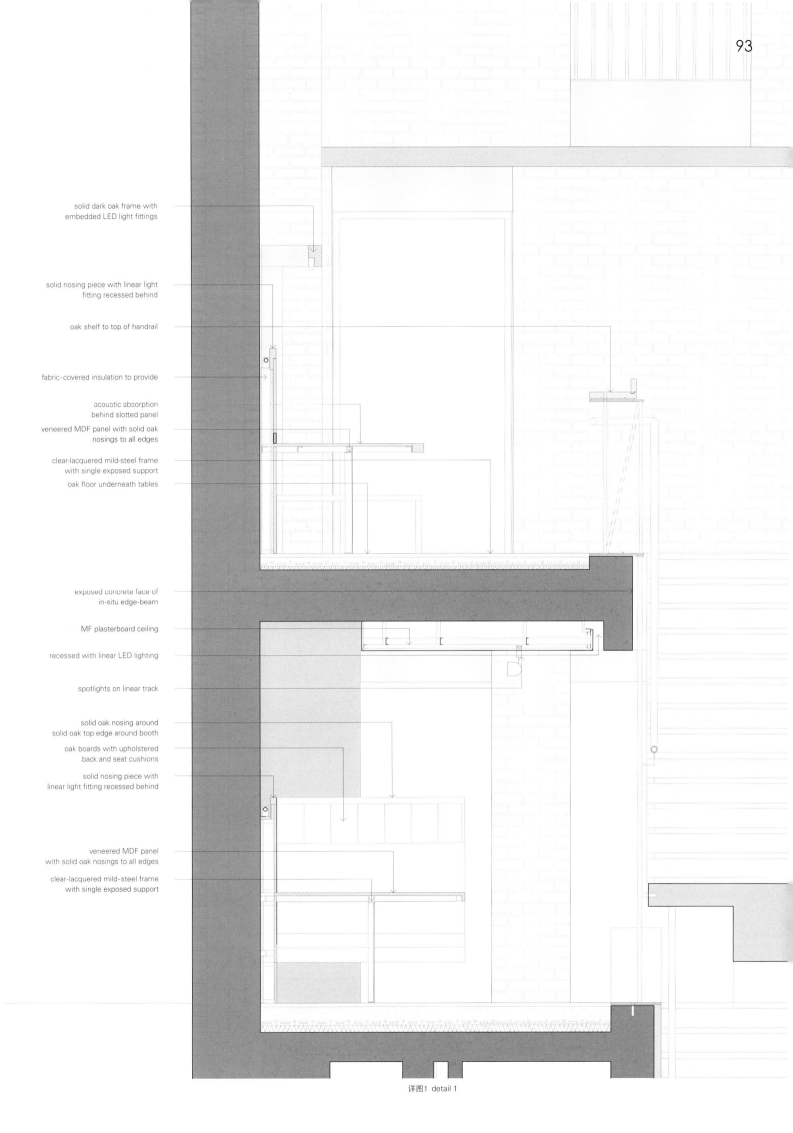

详图1 detail 1

with cellular accommodation; the other is a less regular form with large, individual openings offering into larger, more volumetrically generous spaces.

The foyer occupies the tall voids between these two brick-clad forms and recalls the tight streetscape of the neighborhood – a compressed space characterized by top-light and defined by internal elevations of brick and concrete. Site constraints "forced" the building into the adoption of a stacked section – with galleries on top of the theaters – a kind of "plateau" to which one ascends via the staircases that traverse and ascend their way around the foyer.

The brick cladding of one of the concrete boxes has been peeled away to reveal a 5 stories high concrete wall to the foyer in a very literal revelation of the constructional idea. We proposed that this wall would be finished as a board-marked concrete wall from the earliest stages of the project. As the project developed we considered our ideas in regard to this wall; we recognized that this proposal to use board marking was not driven through a constructional imperative but through an aesthetic instinct to create texture and interest, to relieve that large in-situ surface, and to break it down. This realization led us to consider how we might extend this instinct to a series of "drawings" or "markings" on some of the building's surfaces. Studies of the brick elevations influenced this approach to the stone and concrete walls that have been developed with similar qualities of surface relief. The deliberate introduction of pattern to the concrete wall is expressed as a kind of abstract drawing through the use of smooth concrete framing to fields of highly textured board-marked concrete. This "drawn" or "marked" elevational approach is also employed on the exterior where the basalt surface of the tower presents a surface relief pattern of ribs and recesses to the square.

The MAC is unusual as a new cultural building in the city center in that so much of the interior may be considered as an extension of the public realm. It is in this sense that we hope the MAC is experienced as what we always hoped it would become – a truly public space for people in Belfast. Hackett Hall McKnight

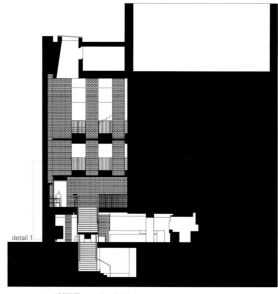

A-A'剖面图 section A-A'

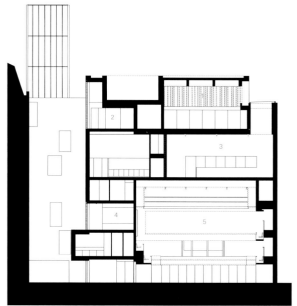

B-B'剖面图 section B-B'

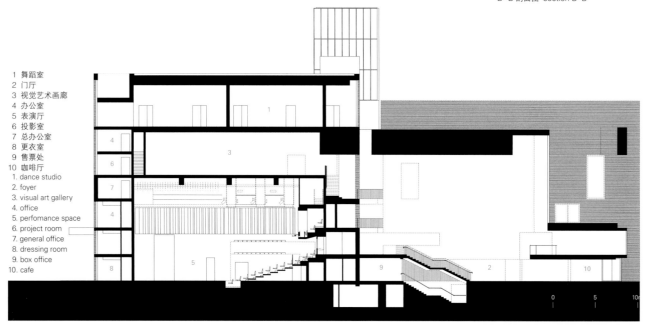

1 舞蹈室
2 门厅
3 视觉艺术画廊
4 办公室
5 表演厅
6 投影室
7 总办公室
8 更衣室
9 售票处
10 咖啡厅
1. dance studio
2. foyer
3. visual art gallery
4. office
5. peformance space
6. project room
7. general office
8. dressing room
9. box office
10. cafe

C-C'剖面图 section C-C'

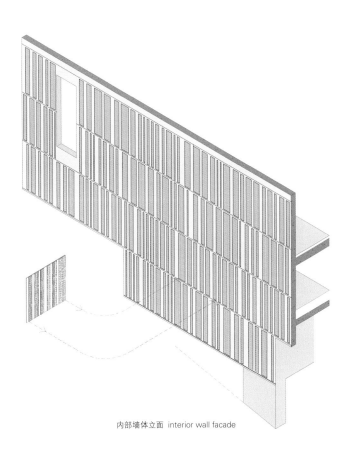

内部墙体立面 interior wall facade

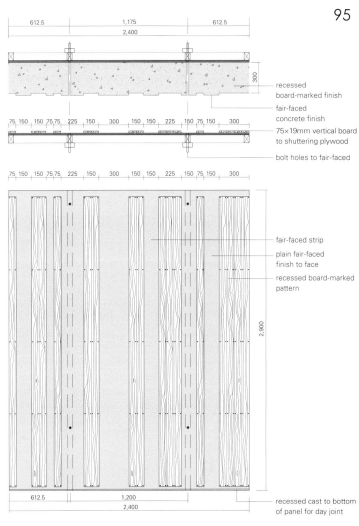

- recessed board-marked finish
- fair-faced concrete finish
- 75×19mm vertical board to shuttering plywood
- bolt holes to fair-faced
- fair-faced strip
- plain fair-faced finish to face
- recessed board-marked pattern
- recessed cast to bottom of panel for day joint

项目名称：MAC, Belfast
地点：Belfast, UK
建筑师：Hackett Hall McKnight
项目管理：Scott Wilson, Braecom(Project Sponsor Client Representative)
结构、机械、电气和维修工程师：Buro Happold
施工技术员：Johnston Houston　剧院顾问：Carr and Angier
声效工程师：Buro Happold　防火工程师：White Young Green
承包商：Bowen Mascott JV　甲方：Metropolitan Arts Center
用地面积：1,500m²　建筑面积：1,500m²　总建筑面积：6,260m²
设计时间：2007　竣工时间：2012
摄影师：©Christian Richters

隆德大教堂广场建筑

Carmen Izquierdo

这座新的大教堂广场建筑位于隆德的中心，直接与大教堂本身相连。场地上还有一座具有历史价值的现存建筑——Arken住宅。

新建筑意在通过适应它周围城市景观的规模和轮廓，从而使自身自然地融入到城市结构中来。与此同时，建筑师的目标是要建造一座现代建筑，在众多历史性的建筑（隆德城市中心环境的特色）中添上一笔。

建筑的外观衍生出了一些新的相邻的公共空间：通向Kyrkogatan大街的入口广场、面向大教堂的入口走廊，以及朝向Kungsgatan大街的三角形广场。除了吸引人的室外空间，建筑师还根据现存建筑和新建筑的形状新建了一个室内门廊和一个室内庭院。

建筑中央的公共空间是从两个入口都可以到达的入口大厅。该大厅是一个大型的公共集会空间，可以举行各种活动，例如招待会、展览以及咖啡聚会。两层高的中庭使阳光可以从顶部射入，同时使公共空间在视觉上与第二层的公共设施整合在一起。礼堂被设计成一个独特的空间，它的天窗向上指向大教堂的塔楼。

建筑的外部简单但是很有特点，它的线条与周围的建筑完美结合。Arken住宅的房檐线在朝向Kyrkogatan大街的一面与入口广场相连。而在朝向大教堂的一面，入口带有极具特色的天窗。

建筑的立面由黄铜合金制成，它是一种天然材料，随着时间的流逝会产生丰富而生动的纹理效果，从而使得该建筑将来变旧时可以融入周围的建筑。刚建成时，它像金子一样闪光，但是几年之后，它就会氧化成暗淡的亚光青铜色。

建筑的内部是用混凝土浇铸成的，呈木板状。内部空间借助光线的灵动性，平衡了这种材料巨大而厚重的特征。

Lund Cathedral Forum

The site of the new cathedral forum is in central Lund, in direct connection to the cathedral itself. On the site is situated the existing Arken house, which is a building of historical value.

The new building aims to integrate itself in the urban fabric in a natural way, by adapting to the scale and lines of the surrounding cityscape. At the same time our vision has been to create a contemporary building that adds a new layer to the many historic layers that characterize the urban environment of central Lund.

The shape of the building creates new adjacent public spaces: the entrance plaza towards Kyrkogatan street, the entrance passage facing the cathedral, and a triangular square towards Kungsgatan street. In addition to the welcoming exterior spaces an internal atrium is created, as well as an interior courtyard, shaped by the existing and the new building.

The central public space in the building is the entrance hall that is reached from both entrances. The entrance hall is formed as a meeting space; a general and generous place which can hold various activities like reception, exhibitions and a cafe. A two-storey

atrium allows daylight to enter from above, while visually integrating the public spaces with the congregational facilities on the second storey. The auditorium is conceived as a unique space, with its skylight pointing up towards the cathedral towers.

The exterior is a simple yet characteristic volumes, its lines playing with the surrounding buildings. Towards Kyrkogatan street the roof lines of the Arken house are continued over the entrance plaza. Towards the cathedral the entrance is signaled by the characteristic skylight.

The facade of the building is made of a brass alloy, a natural material that ages with a rich and living texture, allowing the building to age into its surroundings; at the inauguration it shimmers like gold, but in a couple of years it will have oxidized into a deep and mat bronze color.

The interior of the building is cast in concrete with form of wooden boards. The massive and heavy character of the material is balanced by the play of light in the interior spaces. Carmen Izquierdo

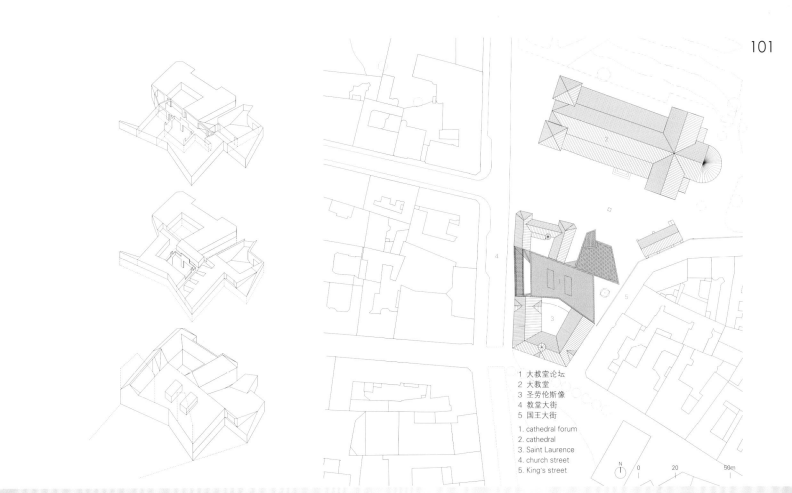

1 大教堂论坛
2 大教堂
3 圣劳伦斯像
4 教堂大街
5 国王大街

1. cathedral forum
2. cathedral
3. Saint Laurence
4. church street
5. King's street

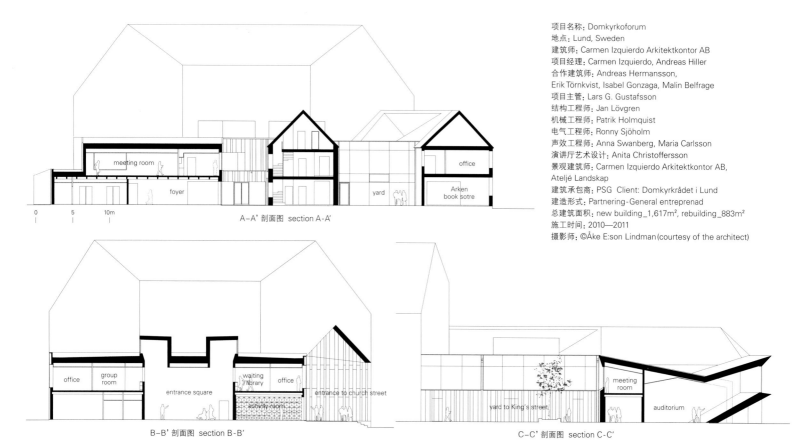

A-A' 剖面图 section A-A'

B-B' 剖面图 section B-B'

C-C' 剖面图 section C-C'

项目名称：Domkyrkoforum
地点：Lund, Sweden
建筑师：Carmen Izquierdo Arkitektkontor AB
项目经理：Carmen Izquierdo, Andreas Hiller
合作建筑师：Andreas Hermansson, Erik Törnkvist, Isabel Gonzaga, Malin Belfrage
项目主管：Lars G. Gustafsson
结构工程师：Jan Lövgren
机械工程师：Patrik Holmquist
电气工程师：Ronny Sjöholm
声效工程师：Anna Swanberg, Maria Carlsson
演讲厅艺术设计：Anita Christoffersson
景观建筑师：Carmen Izquierdo Arkitektkontor AB, Ateljé Landskap
建筑承包商：PSG Client: Domkyrkrådet i Lund
建造形式：Partnering-General entreprenad
总建筑面积：new building_1,617m², rebuilding_883m²
施工时间：2010—2011
摄影师：©Åke E:son Lindman (courtesy of the architect)

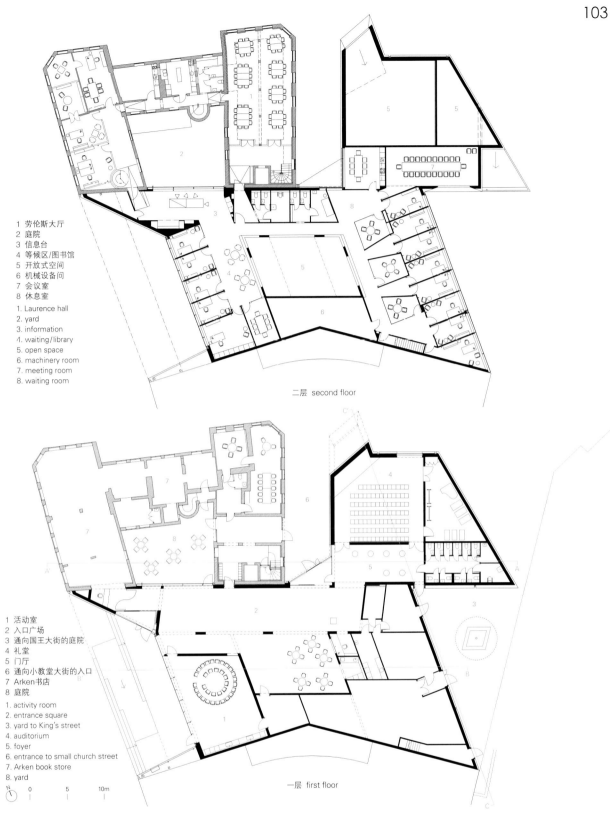

1 劳伦斯大厅
2 庭院
3 信息台
4 等候区/图书馆
5 开放式空间
6 机械设备间
7 会议室
8 休息室

1. Laurence hall
2. yard
3. information
4. waiting/library
5. open space
6. machinery room
7. meeting room
8. waiting room

二层 second floor

1 活动室
2 入口广场
3 通向国王大街的庭院
4 礼堂
5 门厅
6 通向小教堂大街的入口
7 Arken书店
8 庭院

1. activity room
2. entrance square
3. yard to King's street
4. auditorium
5. foyer
6. entrance to small church street
7. Arken book store
8. yard

一层 first floor

圣安东慈善学校修复工程

Gonzalo Moure Architect

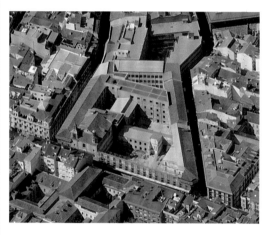

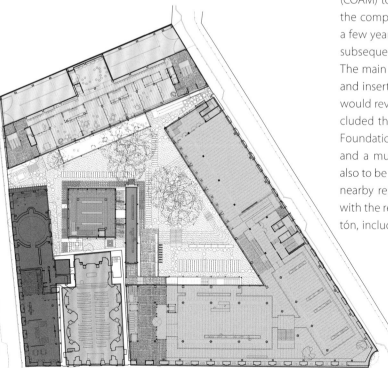

2005年2月15日，马德里市和马德里建筑师协会（COAM）签署了一份协议。协议的主要内容是协调修缮几年前饱受战火折磨的圣安东慈善学校综合建筑的相关事宜，其主要内容包括建筑竞标以及随后中标项目的实施。

此协议的主要目的是推进修缮工作的进行，加快多功能社会公益项目的融入，后者将使慈善学校的城市环境恢复活力。这类项目包括COAM的新总部、COAM文化基金会、托儿所、老年看护中心、运动设施以及一所音乐学校。它还将在地下建造一个大型停车场，可以满足新建筑和周围居民的使用要求。所有这些建设都应该做到不干涉综合建筑内的圣安东教堂的宗教事务，因为它也是修缮工作的一部分。公共可达的开放式绿地作为一种不可或缺的条件，是这座综合建筑的核心。该项目被视为将多样的活动和利益统一在同一建筑体系的机会。

圣安东慈善学校总建筑面积为35 878m^2，其中巴洛克式的圣安东教堂占地628m^2。建筑师Pedro Ribera称它为"文化公益建筑"，社会进步带来的一部分不利影响破坏了该区很多建筑，但是因为圣安东教堂被完全嵌在综合建筑的内部，所以才幸免于难。

这个项目所要追寻的价值就是创造一种宁静但充满活力的氛围。建筑师提出要在保存下来的立面上搭建新的连续平台，这是综合建筑从未有过的。一条凉廊式阳台在白天并不显眼，但在晚上却可转变成活力四射的灯笼，悬置在花园和城市的上空。这是一个让生活变得宁静安详的地方，也是一个让人们重聚的地方。它给整个城市带来了显著的特征，也是一个惊喜，它同时也是一个具有开放性特征但又隐蔽起来的空间；它的道路两旁栽种着树木。在这个高密度的城区中，花园便成了一个开放的地方，让人们可以尽情地呼吸，自然而然地放松下来。这种建筑氛围到处都充满了细声细语的感觉；这里有朴实简单的结构、多利斯式的态度，接近生活本质，物质展现出它的自然价值，并让人深刻理解到"在建筑中，朴实是一种睿智的态度"。它主要是用混凝土、钢材、玻璃以及金色的花岗岩建造成的。

"树荫"在这里指代一处价值连城的空间。城市空间的真正转换就是通过它的建筑和树木来体现的。如果我们能够把城市中带有树木的花园作为遗产留给我们的后代，即是创造了宝贵的城市遗产。

San Antón Charity School Restoration

On February 15, 2005, an agreement was signed between the City of Madrid and the Colegio Oficial de Arquitectos de Madrid (COAM) to coordinate a set of actions towards the restoration of the complex of San Antón charity school that had suffered a fire a few years ago. It included an architectural competition and the subsequent development of the winning project.

The main goal of such agreement was to impulse the restoration and insertion of a mixed-use program of social interest, one that would revitalize charity school's urban context. Such program included the new headquarters for COAM and the COAM Cultural Foundation, a nursery, a day care and senior center, sport facilities and a music school. Extensive underground parking space was also to be built, covering both the needs of the new buildings and nearby residents. All of this should be done without interfering with the religious services that took place in the church of San Antón, included in the complex, and which is object of the restora-

马德里建筑师协会	Madrid architect's association
音乐学校	music school
市政设施	municipal facilities
自助餐厅/书店/音乐室	cafeteria/bookshop/music room
圣安东教堂	San Antón church
书店画廊	bookshop gallery

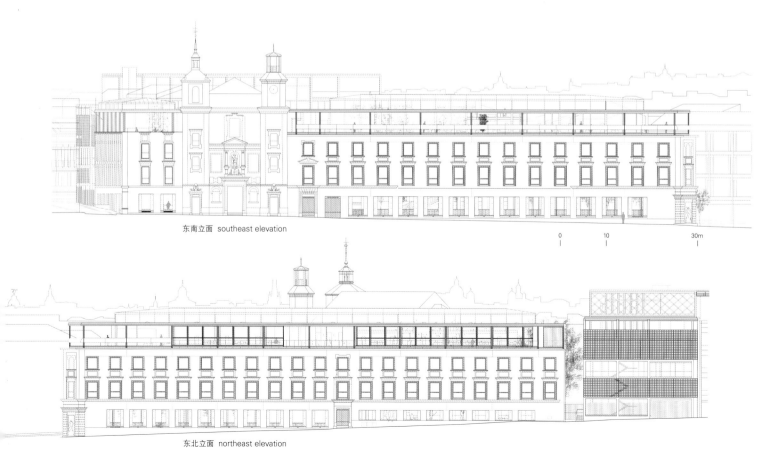

东南立面 southeast elevation

东北立面 northeast elevation

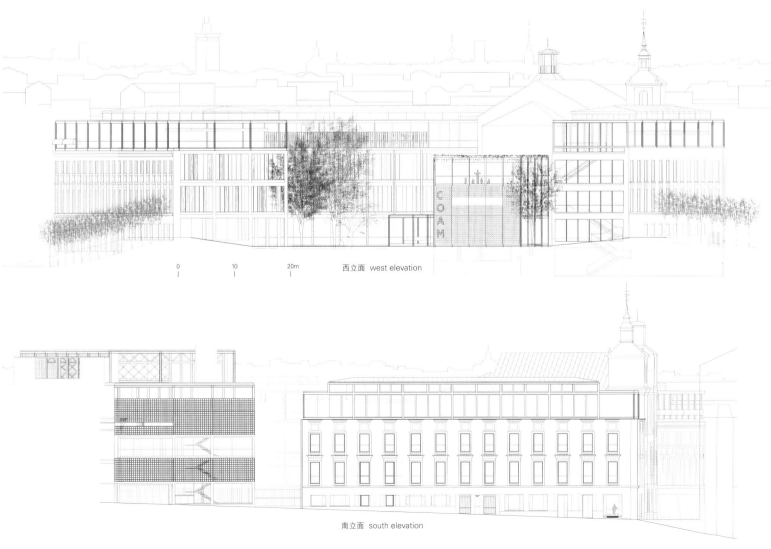

西立面 west elevation

南立面 south elevation

1	主大厅/夹层展厅
2	主楼梯
3	夹层、创新空间
4	花园入口
5	多功能室
6	行政处入口
7	文化公益性质的圣器安置井
8	文化公益性质的外部圣器安置井
9	圣器安置所入口
10	咖啡休息室

1. main lobby, exhibitions mezzanine
2. main stair
3. mezzanine, innovation space
4. garden access
5. multipurpose room
6. administration access
7. sacristy well of cultural interest
8. outer sacristy well of cultural interest
9. sacristy access
10. coffee lounge

一层 first floor

1	永久性展厅
2	马德里建筑师协会画廊
3	礼堂入口
4	礼堂
5	空气处理单元
6	公司转型中心
7	员工区入口
8	厨房
9	主入口
10	员工室
11	更衣室卫生间
12	自助餐厅
13	门厅
14	教室入口
15	打击乐教室
16	薰衣草花园

1. permanent exhibits
2. COAM gallery
3. auditorium access
4. auditorium
5. air treatment unit
6. company transformation center
7. staff area access
8. kitchen
9. general access
10. staff room
11. dressing rooms toilets
12. cafeteria
13. foyer
14. classrooms access
15. percussion classroom
16. lavender garden

地下一层 first floor below ground

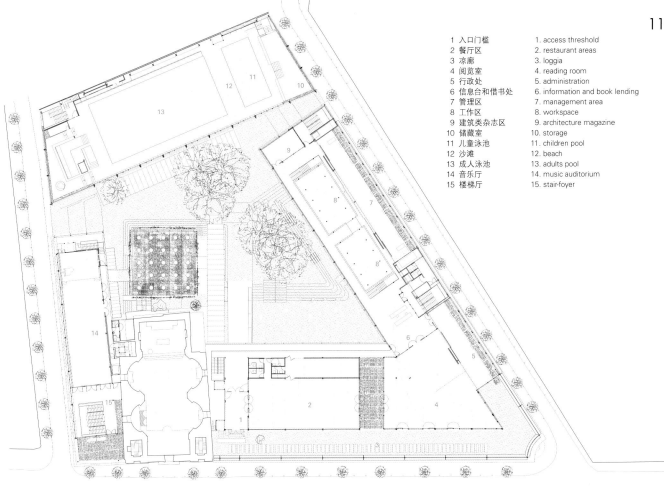

1	入口门槛	1. access threshold
2	餐厅区	2. restaurant areas
3	凉廊	3. loggia
4	阅览室	4. reading room
5	行政处	5. administration
6	信息台和借书处	6. information and book lending
7	管理区	7. management area
8	工作区	8. workspace
9	建筑类杂志区	9. architecture magazine
10	储藏室	10. storage
11	儿童泳池	11. children pool
12	沙滩	12. beach
13	成人泳池	13. adults pool
14	音乐厅	14. music auditorium
15	楼梯厅	15. stair-foyer

四层 fourth floor

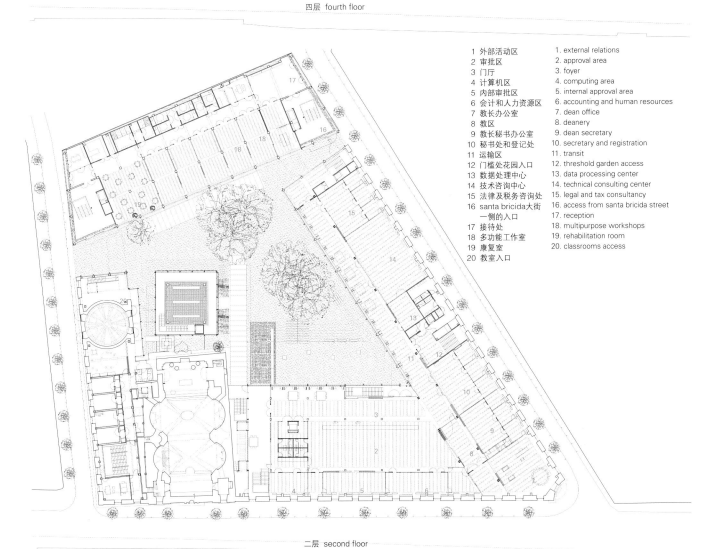

1	外部活动区	1. external relations
2	审批区	2. approval area
3	门厅	3. foyer
4	计算机区	4. computing area
5	内部审批区	5. internal approval area
6	会计和人力资源区	6. accounting and human resources
7	教长办公室	7. dean office
8	教区	8. deanery
9	教长秘书办公室	9. dean secretary
10	秘书处和登记处	10. secretary and registration
11	运输处	11. transit
12	门槛处花园入口	12. threshold garden access
13	数据处理中心	13. data processing center
14	技术咨询中心	14. technical consulting center
15	法律及税务咨询处	15. legal and tax consultancy
16	santa bricida大街一侧的入口	16. access from santa bricida street
17	接待处	17. reception
18	多功能工作室	18. multipurpose workshops
19	康复室	19. rehabilitation room
20	教室入口	20. classrooms access

二层 second floor

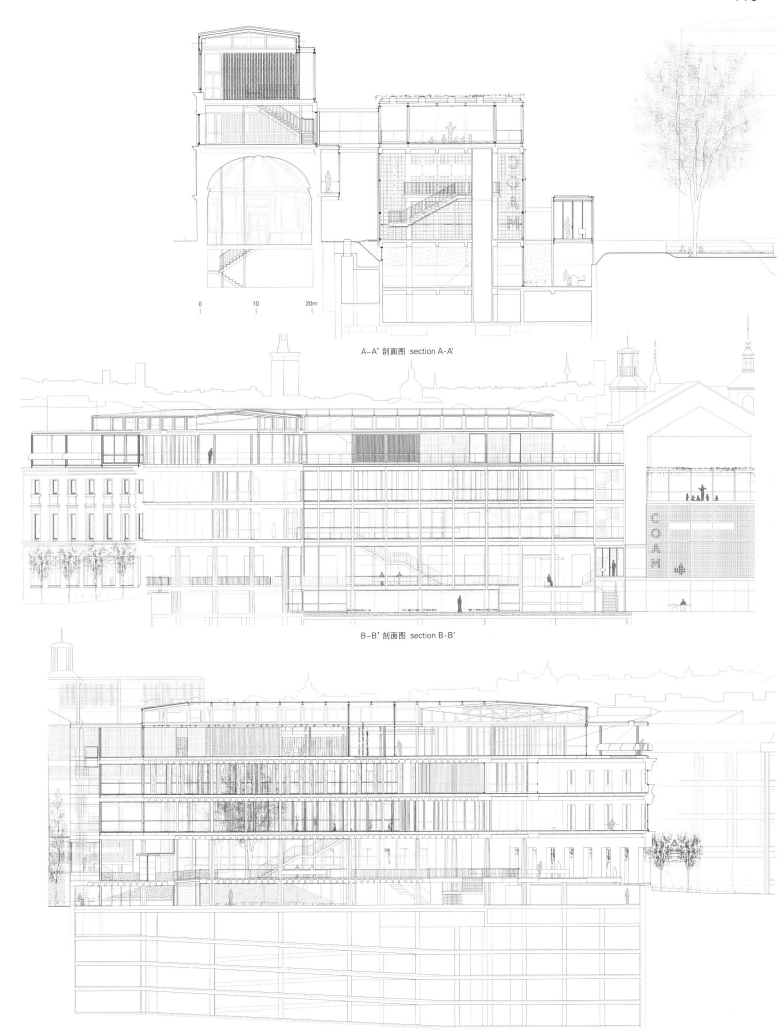

A-A' 剖面图 section A-A'

B-B' 剖面图 section B-B'

C-C' 剖面图 section C-C'

tion as well. As an indispensable condition, a public-access, green open area is set as core of the complex. The program is regarded as an opportunity to consolidate a very diverse array of activities and interests in a single architectonic response.

The total built area for San Antón charity school is 35,878m², of which 628 belong to the baroque church of San Antón, a declared "Building of Cultural Interest" by the architect Pedro Ribera that is completely embedded within the complex and constitutes the only part that did not suffer the progressive ruination that affected the rest of the lot.

The value that the project seeks is to create a serene – yet full of lively contentment – atmosphere. New ledge continuity in the preserved facades is also proposed, as the complex never had, thanks to a loggia-balcony that dematerializes during the day and transforms into a vibrant lantern at night, hovering over the city, above a garden. A place of quietude, of serenity, that embraces life, is a place for reunion. It is also a surprise, an "accent" to the city, a secluded space with an open character, and a place where trees are discovered through transparency at street level. There are a garden in this dense part of the city, and an open space that allows to breathe, where life unwinds naturally. Architectural atmospheres has a whisper-like quality to them, honest and clear construction, a doric attitude, and an approximation to essence where matter reveals its natural values with the understanding that "in architecture, honesty is an intellectual position". It is predominantly built with just concrete, steel, glass and golden granite.

"The shade of a tree" is here regarded as a space of unparalleled value. The true transformation of an urban space is produced both through its buildings as well as its trees. If we are able to pass on a garden with trees in this part of the city as inheritance to our children, we will have built valuable urban heritage.

Gonzalo Moure Architect

项目名称：San Antón Charity School Restoration
地点：Calle Hortaleza 63, Madrid, Spain
建筑师：Gonzalo Moure
项目团队：Myriam Pascual Luján,
Pedro Barranco Vara, Jose María Cristobal González,
Marcos S. Gutiérrez, Fernando Ruiz Martínez,
Enrique Carreras Rufín, David Torres Barrón,
Pablo Matilla Pérez, David Manso Pulido,
Verónica San José González
结构咨询师：Juan Carlos Salva
设备安装咨询师：Ignacio Menendez Azcarraga,
Integral Ingenieria
甲方：Colegio Oficial de Arquitectos de
Madrid(COAM), Ayuntamiento de Madrid
用地面积：7,324m²
总建筑面积：35,878m²
造价：EUR 23,024,717
设计时间：2006
施工时间：2008.3—2012.2
摄影师：©Jorge Crooke Carballal
(courtesy of the architect)

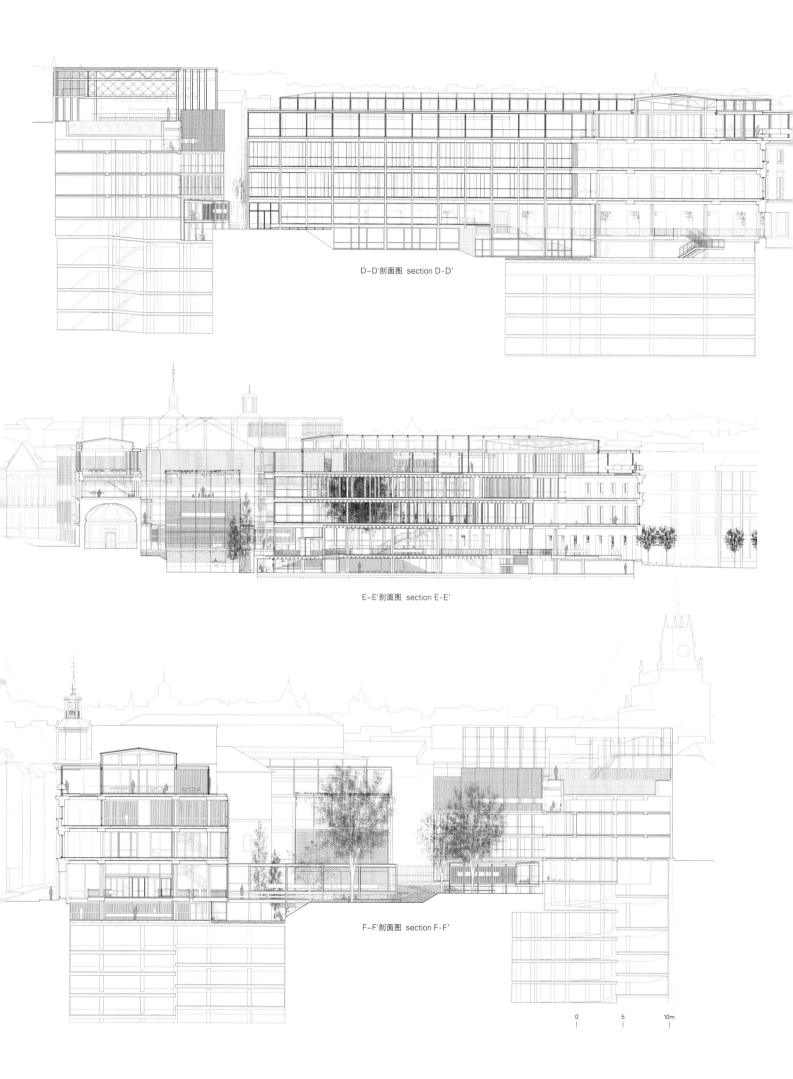

canalón chapa LD 150.90.10

1/2 pie ladrillo
poliestireno extruido

pavimento flotante

grava
capa auxiliar drenante sobre capa filtrante
poliestireno extruido
membrana de pvc con capa geotextil antipunzonamiento superior e inferior

1/2 pie ladrillo

1/2 pie ladrillo

承孝相 Seung, H-Sang

我毫不犹豫地选择了承孝相来代表当今韩国的建筑师。对于他的每一个作品以及他在社会活动中表现出的前后矛盾,舆论褒贬不一,各有所衷;然而,他通过国际间的交流显著提升了韩国建筑在世界范围内的地位,并且通过在国内与其他团队进行跨学科合作,他还推动了韩国本土建筑的发展。因此,承孝相在韩国建筑界是无人匹敌的。他借助直达本质的比喻和简单的语言,使公众更容易理解什么是建筑学,而这一概念却经常被其他建筑师设计的孤高作品所忽略。综上所述,承孝相是一位不仅能运用文字而且总能够以新作品于公众间发出专业声音的建筑师。他为韩国前总统卢武铉设计的墓园给我们提供了一个重新审视自我的机会:什么才是真正的存在?不借助政治议程,纪念仪式的意义何在?又如,Welcome City大厦展示了风景平常的首尔同样有庄严高贵的一面。承孝相对Sujoldang韩屋的重新诠释为建筑形式和传统空间的论战提供了开篇序言;他在Subaekdang项

I don't hesitate to choose Seung, Hchioh Sang as an architect who represents Korea today. There will be preference differences regarding each of his work and discrepancies on his social activities; however, considering how much he has heightened the status of Korean architecture through international exchange and domestically promoted architecture through collaborative works with different disciplines, no architect can match up to Seung. With essential figure of speech and simple language, he facilitates the public's understanding of architecture, which normally had been dismissed in exclusive works by peculiar people. Above all things, Seung is an architect who expertly speaks up by always presenting fresh works, and not just with words. The Graveyard for former president Roh Moohyun provides an opportunity to remind ourselves of true existence and the meaning of commemoration without political agenda; the void created by the Welcome City building demonstrates how Seoul's ordinary scenery could be newly recognized with dignity. Seung's interpretation of Hanok in Sujoldang provided a preamble for a long controversy regarding form and space of inherited tradition and his use of unconstrained design in Subaekdang, shows an understanding of yard-to-room

土地、水、花草、风，360°乡村俱乐部
朝韩非军事区和平生活谷
济州岛艺术别墅社区中心
"MoHeon"与"SaYaWon"

融入自然的小山村/BongHee Jeon

Earth, Water, Flower, Wind, 360° Country Club
Korea DMZ Peace and Life Valley
Jeju Art Villas Community Center
MoHeon and SaYaWon

A Small Village becomes one with Nature/BongHee Jeon

目中自由无束的设计表达出他对"穿庭入室"关系的理解。

  我的个人观点是，承孝相设计的优势在于建筑设计的过程始于对现有空间的研究，基于对使用者的考虑。因此，业主或使用者不会因为不熟悉而产生犹豫，抑或因为建筑的古怪偏颇而顿生迷惑。事实上，原创性和陌生感是展现艺术张力最简单易行的方法。然而，建筑不能只依靠其外表强烈的吸引力，就像人们经过一件艺术品时所产生的视觉冲击一样，因为建筑不只是一种艺术，它还是我们的日常生活。更为重要的是，建筑的终身命运取决于它的地理位置。因此，建筑需要敏感地对场地环境做出回应——它可能会被周围的地势和景观或使用者的生平经历所影响。

融入自然的小山村

  这座村镇是由集聚在一起的民宅合而为一形成的。与我们平常所见的农业村镇一样，村子中央有一棵茂密成荫的大树，周围的几座房屋共同形成了一个大型社区。总体来说，这座建筑含有一个巨大而相互连通的室内空间，但是屋顶和墙体则由若干个小体量组成。每个体量都有一面宽度为7.2m~9.6m的墙。这一宽度比传统韩屋的侧墙要宽，但是又比排成一行的公寓宽度要窄。因此，我们能够根据这些体量识别出房间。此外，这些房间不仅合乎人体尺度，并且空间灵活，可作为办公空间、自助餐厅、更衣室和浴室。

  为了将这些小体量整合到一起成为一个大的室内空间，对连接空间的处理就变得至关重要。对连接空间的处理手法，既要体现出建筑师的创新性，又要体现出建筑的外形。承孝相通过一个室内庭院达到了这一目的。在传统建筑中，庭院是举办各种无法在室内进行的大型活动的场所，或是起到协调各个建筑的连接空间的作用。景福宫中勤政殿的前院，或是有意排在浮石寺无量寿殿前方的安养楼，都起到了这种作用。而在这座建筑中，庭院承担着连接各个体量的责任，或者

relationship.

In my opinion, the advantage of Seung's architecture is that it always starts with us, living right here, right now. Thus the owner or the users don't hesitate from unfamiliarity or be perplexed by its eccentricity. In fact, originality and unfamiliarity are the easiest methods of introducing artistic tension. However, architecture cannot survive solely on appealing with a strong stimulus that you normally get when you pass by an artwork because architecture is not only art but also our daily life. More than anything, the fate of architecture is permanently determined by its location. For this reason, architecture has to respond sensitively to the character of the site, which could be influenced from the surrounding topography and landscape or the lifetime and experience of the users.

A Small Village becomes one with Nature

This house is a town composed of several densely gathered houses. As we normally see from an agricultural town, there is a huge shade tree in the center and little houses come together to form a big community. Overall, this house holds one huge connected interior space but the roof and the walls are composed of several smaller masses. Each mass has a side with the width of only 7.2m to 9.6m. Such width is wider than the width of Hanok's side wall but narrower than the width of the apartment aligned in one line. Therefore, we recognize those masses as houses. Even more, the dimension is closely set to human body and motion and is adapted from the measurements from the office, cafeteria, locker room, and bathroom.

In order for these small masses to come together to form one big interior space, dealing with the areas of connection is crucial. From the way these connections are treated, an architect's creativity and the building's formal achievement are recognized. For this purpose, Seung utilized an interiorized yard. In traditional architecture, a yard serves as a space for huge events that can't be held in a building or be a means to harmonize building connections. Such role of the yard can be seen from the wide yard in front of Geun-Jung-Jeon in Gyeongbok palace or the placement of Ahn-Yang-Roo in front of Moo-Ryang-Soo-Jeon in Bu-Suk-Sa. In this house, the yard deals with connecting each masses or accommodating large space that can't be solved with a narrow elongated

从外部看是几座小房子组成的一个大村镇。
The exterior showing several small houses forming one big town.

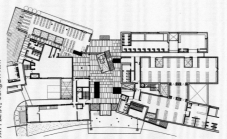
将主厅作为村中虚构的庭院,若干房间围绕在其周围。
Treating the main hall as an imaginary town yard, several houses are assembled around the parimeter.

说它提供了一个窄长体量无法实现的敞亮的大空间。唯一的区别就是这个院子是个带顶的室内空间。

整个室内空间的中心是中央大堂,其中还设有一个光照充足的庭院。如果站在这里环顾一周,就能够看到两侧各排了三四间房子,仿佛翅膀一样铺展开来。庭院和入口稍有距离,可将光线引向周边暗区;因此显露出其原本是一个介于住宅之间的室外庭院;之所以稍有角度是为了将人流自然引向餐厅。中央大堂扮演了这座村镇的中央庭院的角色。将餐厅和厨房分隔开的过道以及通向男更衣间的小径会让人联想到村中的胡同。同时,建筑师还在这里设计了一个小院,给人一种在村中漫步的感觉。因此这座建筑的用户不会感觉到他们生活在室内,而更像是穿庭过巷到每个房屋。为达到此目的,所有过道的墙,即便是内墙,也都像外墙一样做成粗糙的混凝土表面,并且地面也都由具有粗糙表面的花岗岩铺就而成。

如果说中央大堂和庭院平衡了室内空间的布局方式,那么带有设备井的楼梯间则从外观形式上建立起了整座建筑的竖向中心轴线。建筑外形,跟室内空间一样,能让人们联想到那种由众多狭长窄仄的房屋(其人字屋顶相互连接)组成的小镇。这些房子彼此间稍成角度聚集在一起,看起来更为自然和谐。相比建筑中其他横向排成一长列的形式,入口接待台后面的楼梯间则要高出一层,以便于人们去往屋顶,而高出屋顶的室外构件则使得楼梯间从形式上看起来像是整座建筑的一个塔楼。这个竖向的建筑体量使水平延展的室外空间得到平衡。

建筑师在入口的雨篷处重点强调了水平线条。雨篷高度勉强高过头顶,这给人的第一印象与其他高尔夫俱乐部大不一样。高尔夫在韩国仍被认为是上层阶级的运动,因此高尔夫俱乐部的入口规模巨大、形式庄重,往往可以媲美高级酒店的入口。雨篷是这样一种装置:它会让到这来的人产生自命不凡的非常态心理,或体验一种与以往全然

mass. The only difference here is that the yard is an interior space with a roof.

The center of the entire interior space is the central lobby with fully lit courtyard. If you stand here and look around, three or four houses are arranged side to side like wings that are spread apart. The courtyard that is positioned a bit off from the entrance brings in light to otherwise stifling interior, thereby reminding that it is originally an exterior yard in between the houses. The reason why the angle is skewed is to naturally guide the traffic from the entrance to the cafeteria. The central lobby acts as the central yard of the town. The passage that divides the cafeteria and the kitchen and the path to the men's locker room evoke a town alleyway. Also, a small yard is introduced in the intervals to give an impression of walking through a town. So the people who use this building don't feel like they are inside of a building but rather experience passing through yards and alleys into each house. For this purpose, the walls of an alleyway, despite the fact that it is an inner wall, are treated with rough concrete finish like an outer wall and the floor is granite with rough finish.

If the central lobby and the courtyard balance the composition of the interior space, the staircase with facility shaft sets up the vertical central axis in the overall composition of the exterior form. The exterior form, like the interior spaces, reminds one of a town composed of long, narrow houses with connected gable roofs. The houses are clustered in slight angles that they look even more natural together. Compared to other forms that are arranged in a long line, the staircase behind the entrance lobby counter has to go up one more floor to access the roof and the external unit is placed on top making it look like a tower that stands high above all the other forms. This vertical mass balances out the horizontal exterior.

The emphasis on horizontal line is reiterated in the canopy of the entrance. The canopy that barely passes human height provides the first impression that is very different from the ordinary golf clubhouse. Golf in Korea, is still identified as a sport for an upper class, so the entrance of a clubhouse often emulates high society hotel lobby with exaggerated size and imposing shape. It is a device to induce atypical conceit or fresh experience from the

更衣室中的小型庭院对空间起到了划分作用,同时也使室内更加明亮。

Little courtyards in the locker room divide up the space and make the space brighter.

不同的感受。因此这种俱乐部也愿意以庄重的异域造型告诉公众它来自不同的时空。然而,承孝相却选择使用完全不同的方式来实现他的设计。它既不具有异国风情,也没有因为刻意模仿遗失的过去而丧失原创性;它使用的都是非常普通平常的元素。通过有条不紊地组织日常建筑语汇,它使平凡的事物也可以再现高雅庄重之感。

对于细节的处理是使平凡事物上档次的先决条件。例如,暴露在外的表面粗糙的混凝土墙,上面排满了一种特殊的松木,看起来好像一块块砖叠摆在一起。又如中央大堂的天花板,匆匆一瞥看似粗糙简陋,但上面的灯具需要很复杂的铰接安装。露梁天花板看起来简单易行,然而装在上面的灯具为了制造出自然光的效果必须加盖一层半透明罩。再如,其他空间将自然光引入以增强空间立体感。

建筑师希望室内空间室外化的想法还表现在对地面的处理上。从入口接待处到中央大堂,再到窄径和球场上,所到之处皆均匀地铺设了花岗岩饰面。地面是决定空间整体氛围的一个基本因素,同时它也是建筑中唯一会跟人体产生接触的部分。不仅如此,对于始终保有席地而坐这一习俗的韩国人来说,地板更是最熟悉的建筑元素。即便不易于清理和维护,花岗岩地面也一直从主入口延伸到球场入口——这是建筑对我们身体的敏感性和记忆力进行全面解读后所呈现的空间效果。如果你的观察力足够敏锐的话,就会发现花岗岩地面最终消失于与自然接壤的地方。

客人进入俱乐部之后,将背包留在前台,经过简单的流程,转到另一辆装着他背包的车上,然后驶往另一个地点,这一过程与机场类似。尚有一点不同就是更衣的过程。旅途中的更衣可以比作是单调乏味的安检和入境检查。跟机场一样,俱乐部也有两个入口。一个通往日常生活区,另一个与非日常生活区相连。来打球的人可以从前者进入,而尽兴而归的人们可以从后者离开。所有的门本质上无外乎分为

people who enter. Thus the golf clubhouse enjoys imposing exoticism to the public that it came from somewhere else from some other time. However, this house uses a completely different approach. It is not foreign nor it imitates the lost past that it doesn't allow originality; it uses quite ordinary elements. By systematically composing everyday architectural language, it revives the dignity of the ordinary.

The handling of the detail is a prerequisite to give class to the ordinary form. Especially the rough sides of the exposed concrete wall use a special pine tree cast that gives an impression of stacked bricks. Also, the ceiling of the central lobby at a glance appears to be crude, but it is treated with very articulated lights. The ceiling with exposed supporting beams could at first seem unsophisticated; however, the installed lights have translucent cover to get an effect of natural light. Also, for other spaces, natural light is brought in to augment three-dimensionality.

An effort to make interior space appear to be exterior continues on the floor. The floor from the front reception to the central lobby, then to the path and field leading into each space is uniformly treated with granite finish. A floor is a fundamental means to determine overall atmosphere of the space and the only part of architecture that makes direct contact with human body. Moreover, for Koreans who maintain customs of sitting on the floor, it is the most familiar architecture. Despite the difficulty of cleaning and maintenance, the fact that the floor pattern is retained from the front entrance through the field entrance shows full understanding of our body's sensibility and memory. A astute person would realize the granite disappearing and being dispersed into nature. The clubhouse where one leaves one's bags at the front upon arriving, goes through simple procedures, transfers to another car where one's bags are loaded, and departs for another location, is similar to an airport. One thing that is different would be that one changes clothes in the meantime. The changing of clothes for a journey could be compared to a tedious process of security check and immigration inspection. Just like an airport, the clubhouse consists of two entrances. One is a passage that is connected to the daily life and the other is one that is connected to the non-daily life. The former is for a fresh start and the latter is for the return. De-

面向球场一侧的入口门前铺有人造石面砖,铺地边缘逐渐消融于自然。
The artificial rock floor pattern in front of the entrance facing the field gradually scatters apart towards the nature.

照片提供:JongOh Kim

"出发"和"归来"两种。因此,机场和俱乐部也可以被看作是介于前门和后门之间的空间。

人们对俱乐部的前后立面抱有不同的期望。如果说作为全新开始的正面入口必须引起游客的期盼心理和神秘感——通过将现有环境以外的未知世界罩上面纱的方式,那么作为回归的背面入口则必须温暖地欢迎旅途归来的人们。虽然两个入口看似一样欢迎来客,但是前入口的墙面上设有小窗,而面向球场的入口立面则全是大型窗户;因此我们能够辨认出这才是真正的前立面。从建筑外部看感觉无法进入内部,但是从球场一侧看过去,令人惊奇的是俱乐部的形状被整体拉长了。这是因为行政部门用房的辅助设施大规模地集中在这一侧,并且俱乐部还在此设置了背景屏幕,该屏幕遮盖住了墙上的四个洞口。因此,俱乐部的前立面到底是哪边就显而易见了。外部入口只是进入俱乐部的一个通道,通过它你可以斜着看到立面;而当你走向内部入口时则会看到进出球场的人们,迎面就能进入场地。考虑到外部入口只面向停车场,而内部入口却可坐拥绿地、主高尔夫球场,所以这种安排是符合逻辑的。因此,俱乐部占据了大量的南侧用地,挡住了球场的整个北侧用地。

360°乡村俱乐部是一块开放的公共场地,不分会员级别,非会员也可使用。大多数来此的人们都是利用周末时间来玩高尔夫的。他们通过运动与自然相结合的原始方式,远离城市中浮华喧嚣的日常生活。这样看来,理智的做法是以尽量少的人为方式为此类人群提供一处休息舒缓的去处,而非用吵闹的集会或自以为是地鼓吹与城市酒店的合作。承孝相认为穷人的审美正源于此逻辑。他解释说这里所说的穷人并不是真正的穷,而是自己愿意受穷。就针对这一观点的不同意见来说,他的说法多少有些主观臆断。这段话会让人联想到圣经里对这种内心贫穷的人的描述。考虑到承孝相是一个虔诚的基督教徒,他

parting and returning are indeed two intrinsic qualities of all gates. Thus, the airport or the clubhouse could be viewed as a space in between the front and back gates.

Each front and back facade of the clubhouse has dissimilar expectations. If the front entrance for a fresh start has to generate anticipation and mystery through veiling the unknown world existing beyond, the back entrance for the return has to have warmth welcoming those who are coming back from their journey. Even though both entrances have similar appearances of open arms, the front entrance is enclosed by walls with small windows whereas the entrance facing the field is full of wide windows; so we can tell that this is the true front. It is hard to perceive approaching from the outside, but from the field, the overall form of the clubhouse is surprisingly elongated. This is because the facilities used by the administration division are connected extensively to the side, and with it, the clubhouse made a backdrop screen that blocks 4 holes. Subsequently, it becomes apparent where of the clubhouse is the front. The outer entrance that one faces as one approaches towards the clubhouse provides only a slanted view of the facade, however, when approaching the inner entrance that one sees going out and coming back from the field, one approaches head on. Considering the fact that the outer entrance only faces the parking lot whereas the inner entrance shares the field, the main part a golf course, such setup is logical. Accordingly, the clubhouse occupies a large side of the south obstructing the entire north side of the field.

360° country club is a public course. It is an open space that can be used by anyone with no classification of members or non-members. Most of the people who come to this course are weekend golfers who desire to get away from their artificial business-like daily life through the most primal combination of nature and exercise. If so, it is rational to provide a rest and comfort in nature with limited artificiality instead of noisy gatherings or conceited boasting associated with urban hotels. A poor man's esthetics, acknowledged by Seung, is related to this logic. He explains that a poor man here is not a person who is poor but a person who is willing to be poor. It is a sort of self-assertion to different speculations surrounding this notion. The phrase reminds one of

从高尔夫球场看向会所。整个北面形成一个"欢迎归来"的大背景墙。
The clubhouse viewed from the field. Forming a backdrop blocking all of the north side, it welcomes the return.

照片提供：JongOh Kim

很可能是引用圣经，但这一点并未得到证实。无论如何，在当今这个物质与效率当先的社会能思考内心贫乏问题还是难能可贵的，其高尚之处在于它将建筑赋予了一层精神意义。尤其是，在穷人迫切的希望与需求中，我们能够发现一种悲剧美。

从该建筑不带有任何臆想与人工色彩这一点，以及它对真实、自然和原始意向的凸显，不难看出这确实是为内心贫苦的人建造的房子。每个小体量都是一个装饰极简的立方体，甚至各个立面也都如盒子一样交接在一起。墙面、地面和天花板都没有粉刷，将其未经处理的样子保留下来。唯一一处装修华丽的就是建筑各体量之间的小院子，里面充满了阳光、微风和绿地。墙上光影移转，地上满是野生植物的落叶，透过窗户望去是满眼的蓝天、高山和田野。在这座建筑中，一个人能够彻底地降低自己抬高别人，学会去仰视周围事物。高尔夫球场中央的"影之屋"也是如此。虽然形式和材料不同于过去，但是它的选址方式似乎与传统的观景台极为相似。与观景台一样，该房屋只提供坐席，让坐下停歇的人可以将风景尽收眼底。虽然只作片刻停留，但透过卫生间的天花板，人们可以仰望蓝天，这让人们感觉到即使在最私密的地方自然也触手可及。承孝相设计的无顶篷式卫生间最先见于他在Su-Bak-Dang的设计中，而现在这已经成为了他的惯用手法。

作为一个邂逅和分离的地方，是否应该将机械化的灵敏高效性作为机场中最重要的因素在建筑界始终存在争议。无论机场还是俱乐部会所，都是一个人来人往、有例行流程和规定的地方，因此都面临着相同的难题。这座建筑将各个元素巧妙地组合在一起，不缺不过，其中包含的精神美学让人在此能够充盈内心的空虚，并通过自省面对现实。它之所以会成为一座优秀的建筑就是因为它并没有仅仅满足于简单的设计，还包含了建筑师精心设计的高效性和道德性。

a person poor at heart described in the Bible. Considering the fact that Seung is a devout Christian, it is possible that the Bible is the reference, but it is not confirmed. Regardless, thinking about poverty in today's age of materialism and efficiency is noble in that it brings out a spiritual element in architecture. Especially, there is certain tragic beauty that can be detected from the poor's desperate hope and urgent need.

The fact that this house is for the people who are poor at heart is evident from restraining all conceit and artificiality to reinforce truth, nature, and original intention. Every mass is a cube with the least amount of ornamentation and even the sides are assembled as a box. Sides that form the walls, floor, and ceiling are not painted and are left unprocessed. Things that are ornate in this house are the small yards punctured in between the masses and the light, wind, and grass inside. There is a movement of shadows on the walls; leaves of the wild plant on the ground; and the window filled with sky, mountain, and field. This house is like a stand where one thoroughly lowers oneself to boost the others. The house of shadows in the middle of the golf course does the same. The form and the material are different, but the way it is situated seems exactly like a traditional gazebo. Same as the gazebo, the house only provides places to sit, and the person who sits down is the one who looks upon the scenery. Although for a brief moment, the sky viewed through the ceiling from the bathroom shows that nature could be attained even from the most private space. He has previously introduced a bathroom with open sky in Su-Bak-Dang and now, it has become his manière.

Should the machine-like efficiency or sensibility as a place of encounter and separation be put above all things in an airport had been long debated among architects. An airport and a clubhouse, both being places of departure and arrival with fixed procedures and customs, are facing the same dilemma. This house, delicately composed of no excess or shortage, is equipped with spiritual aesthetics where one fills up through emptiness and faces the reality through introspection. The reason why this house becomes a good architecture is that it doesn't simply end with simple design but it consists of elaborate efficiency and ethicality. BongHee Jeon

土、水、花、风
360° 乡村俱乐部

当高尔夫球场有着"土、水、花、风"这一特殊的名字时，它便显得更加贴近大自然。自然在这里意味着源泉，重新唤醒了在城市和平常生活中耗尽的能量。当然这不只意味着物质上的重启。这种只能在自然中找到而不存在于日常生活中的美——泥土的味道、水流的声音、鲜花的芳香以及风的清新对我们来说都很新鲜，让我们更坚信我们的世界非常美丽，这也激发出我们对生活的自豪感。这可能就是高尔夫所蕴含的真正的精神。

俱乐部会所是球场的转折点。这是一条从城市到自然的通道，同时又为人们提供了一个从日常生活进入非日常生活的空间。因此，这两种截点在那里共存并创造了一个让二者发生戏剧性转换的空间。此外，这个俱乐部不是针对特定人群的封闭空间。这个场所能让许多互不熟悉的人在同一时间聚集在这里，是一个重要的公共社区。这时，容纳这些活动的建筑已经不只是一座建筑，而更像是一座城镇或者城市。也正因为这个原因，俱乐部会所被按照城镇来设计。

这个由很多房子组成的会所同时也有另一个作用，就是要能完全满足功能需求以便更活跃地操作这个系统。换句话说，每一个对应特定功能的构件都被预先设计好并与其他构件紧密连接。正因如此，它们以一种不规则的形式聚集在一起，并且这种不规则形成了一个自然的社区形式。此外，由于个体单元的群聚，介于这些建筑之间的空间有时会形成一个庭院，创造出特别丰富的视觉效果、自然通风以及光照条件。

这些体量大部分都只有一到两层，它们以其复折式屋顶和整体的斜屋顶创造了一种美丽的景观，就像一个小村庄或是山中的寺院，所有这些也加强了这组建筑群的可识别性。

混凝土和石材是建筑外表面主要应用的建筑材料，用这些来预示建筑可以完全接受大自然的美丽变化。门窗的木框连同木材的表面都暴露在外，以此来暗示建筑内部的舒适。

从外部驶进来的车辆会感觉好像进入了一个城镇。一个与俱乐部会所相连的宾馆环抱着接待区，更增强了这种感觉。这些循环通道被集约化处理以便那些来迟了的高尔夫运动者能快速做好准备上场。

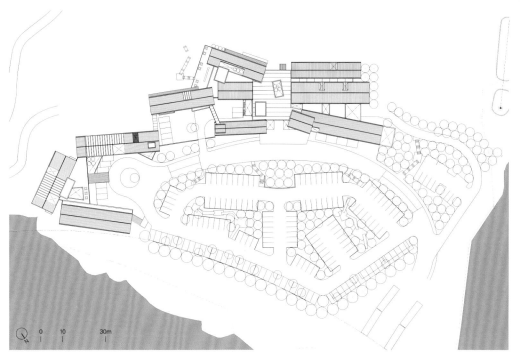

然而，对于那些时间充裕的人来说，在更衣室外的庭院周围提供了几处可以自如走动的空间。

站在旷野上的最后一个大厅中向外看，小镇建筑群的钛合金屋顶都闪闪发光，像是在迎接高尔夫运动者。从浴室看刚才他们经过的旷野有一种看到记忆中的风景的感觉，同样，在餐厅、宴会厅或室外露台看到的也都是这样的美景。

对于每一个在这里结束整段旅程回到日常生活的人来说，俱乐部会所就像一座由土、水、花、风构成的记忆之城，所以，人们在这里获得力量，又重新回到了日常生活中。

Earth, Water, Flower, Wind, 360° Country Club

Earth, Water, Flower, Wind, the golf course with the special name means nature. Nature here has the meaning of the origin which reboosts the power exhausted in the city and the daily life. It of course doesn't mean the physical reboosting only. The beauty of non-daily life found in nature, the smell of earth, sound of water, the fragrance of flowers, and the freshness of wind come new to us, confirming how beautiful our world is, and encourage the pride to our life. That's probably the true spirit of golf.

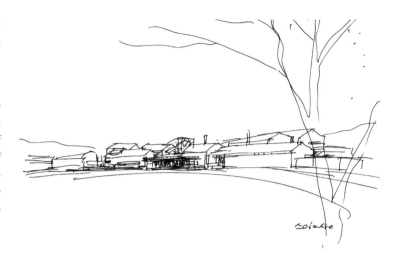

西北立面 northwest elevation

东北立面 northeast elevation

东南立面 southeast elevation

西南立面 southwest elevation

1 自助餐厅 2 室内球场 3 等候室 4 员工休息室 5 厨房 6 入口大厅 7 办公室 8 高尔夫车存放处
1. cafeteria 2. inner court 3. waiting room 4. staff room 5. kitchen 6. entrance hall 7. office 8. cart parking

A-A' 剖面图 section A-A'

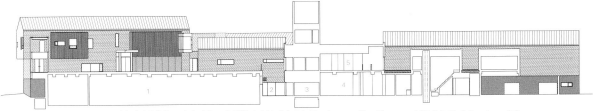

1 高尔夫车存放处 2 室内球场 3 洗手间 4 大厅 5 入口大厅   1. cart parking 2. inner court 3. toilet 4. hall 5. entrance hall

B-B' 剖面图 section B-B'

1 机房 2 办公室 3 洗手间 4 储藏室 5 休息室 6 球具专卖店 7 自助餐厅 8 空调设备间
1. machine room 2. office 3. toilet 4. storage 5. lounge 6. pro shop 7. cafeteria 8. airconditioning unit

C-C' 剖面图 section C-C'

Club House is its turning point. It is a gateway from the city to nature and a space changing from daily life to a non-daily one. Thus, the two cut-off points co-exist there and create a dramatic transference of the space. Moreover, the club is not a closed space for certain people. The place where many unfamiliar people come together at the same time is an important community and the space that puts it is not a building but already a town and a city. For that reason, Club House is designed to be a town.

Club House which is composed of many houses also has a purpose to completely carry out functional requirements to operate the system busily. In other words, each volume that fits for an individual functional unit is postulated and is later connected to each other closely. Because of this, they are gathered in an irregular form and such irregularity created a natural form of a community. Besides, the spaces between created as a result of the clustering of the individual unit become a courtyard from time to time and provide a special richness visually as well as natural ventilation and lighting.

The volume which mostly consists one or two stories strengthens the recognition of the house with its gambrel roof and the collective figure made by the pitched roof creates a beautiful scenery just like a small village or a mountain temple.

For external facing, concrete and stones are used as main materials to signify the background and then fully accept the beautiful change of nature. The wood around the opening is exposed together with the surface of the wood in order to imply the coziness of the inside.

The vehicles accessing will feel like they are entering a town. A guesthouse connected with the Club House embraces the access area and enhances the feeling. The circulation is minimized for the golfers who may arrive the town late and hurry to start teeing up quickly. However, for anyone who has enough time, the locker room offers easy movements surrounding the courtyards at several places.

Standing at the last hall in the field, the titanium roofs of the town give off a shine and greet the golfers. The fields that they just passed by or are in the bathroom are the sceneries of memories, and the passing lights seen in the dining room, the banquet hall, or outside terrace occasionally.

Club House remains as a town of memories created by earth, water, flower, and wind for everyone who goes back to their ordinary life after finishing the whole courses. So they gain the power here to start the daily life again. Seung, H-Sang

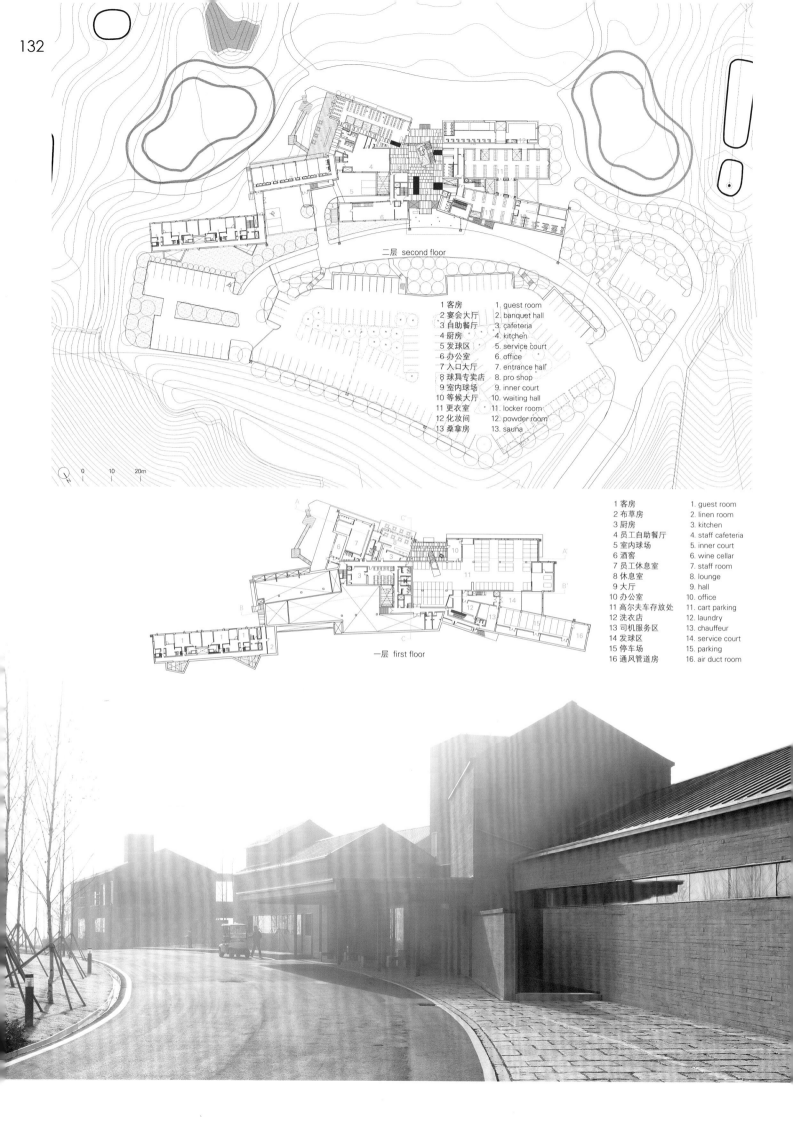

二层 second floor

1 客房	1. guest room
2 宴会大厅	2. banquet hall
3 自助餐厅	3. cafeteria
4 厨房	4. kitchen
5 发球区	5. service court
6 办公室	6. office
7 入口大厅	7. entrance hall
8 球具专卖店	8. pro shop
9 室内球场	9. inner court
10 等候大厅	10. waiting hall
11 更衣室	11. locker room
12 化妆间	12. powder room
13 桑拿房	13. sauna

一层 first floor

1 客房	1. guest room
2 布草房	2. linen room
3 厨房	3. kitchen
4 员工自助餐厅	4. staff cafeteria
5 室内球场	5. inner court
6 酒窖	6. wine cellar
7 员工休息室	7. staff room
8 休息室	8. lounge
9 大厅	9. hall
10 办公室	10. office
11 高尔夫车存放处	11. cart parking
12 洗衣店	12. laundry
13 司机服务区	13. chauffeur
14 发球区	14. service court
15 停车场	15. parking
16 通风管道房	16. air duct room

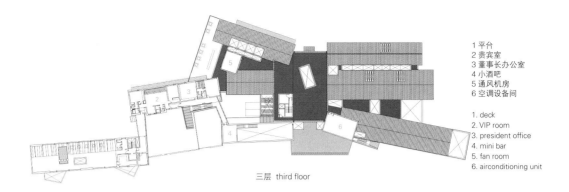

1 平台	1. deck
2 贵宾室	2. VIP room
3 董事长办公室	3. president office
4 小酒吧	4. mini bar
5 通风机房	5. fan room
6 空调设备间	6. airconditioning unit

三层 third floor

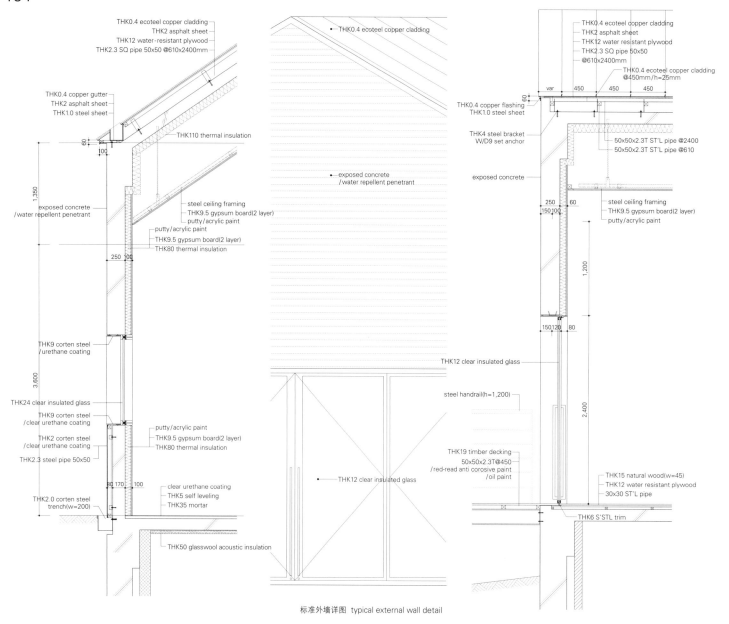

标准外墙详图 typical external wall detail

项目名称：Earth, Water, Flower, Wind, 360° Country Club
地点：GangCheon-myeon, YeoJu-gun, GyeongGi-do, Korea
建筑师：Seung, H-Sang
项目团队：JongTae Yun
结构工程师：Seoul Structural Eng.
机械工程师：Seah Eng.
电气工程师：URim Elec.
景观设计师：Seoahn Total Landscape
用地面积：818,806m²
建筑面积：4,999.04m²
总建筑面积：9,566.81m²
设计时间：2009.1~2009.10
施工时间：2009.10~2011.8
摄影师：©JongOh Kim

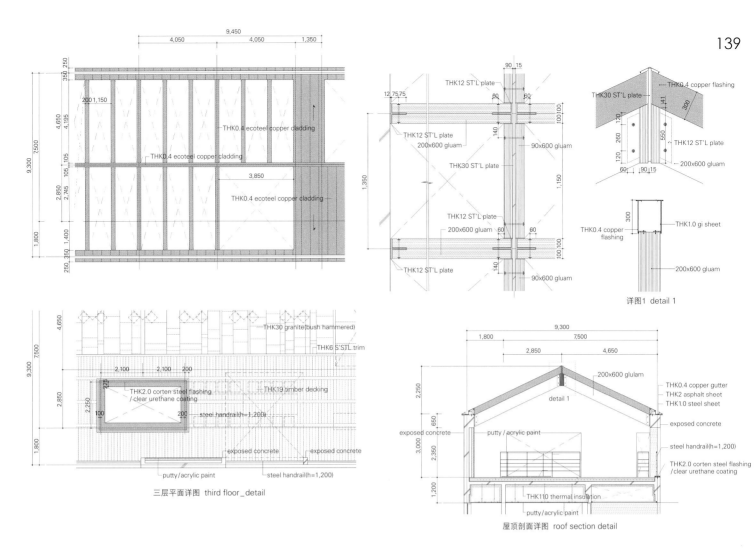

三层平面详图 third floor_detail

详图1 detail 1

屋顶剖面详图 roof section detail

朝韩非军事区和平生活谷

大地上的建筑

这个有着异乎寻常的名字的地方——朝韩非军事区和平生活谷——隶属于一个口号是"用生活的钥匙打开和平的大门"的组织。值得注意的是,非军事区(DMZ)这一战争的产物究竟如何成了一个恢复生产生活的场所?这个组织正是因此而成立的,旨在让人们思索当今时代,同时保护生命的神圣和尊严。江原道的巨大资金支持和致力于这项事业的学者们使得这个组织能够顺利成立。这些学者也是该组织的主要成员。

这一设施的基本用途是为DMZ研究提供基地,但是这里也组织演讲、研讨会和冥想,还包括一项促进恢复生产生活的农业活动。正是因为这样,设施本身需要和自然保持紧密的关系。从一开始,这个项目就不允许破坏或控制自然。

建筑地点位于江原道麟蹄郡的一座山上,距离DMZ很近。半山腰有一块朝西的台地,连接着绵延的山脉和车道,因此是一个至关重要的元素。建筑在此是一个连接各种元素的媒介,所以必须在自然和人工之间达到很好的平衡。如果从山顶鸟瞰,大地好像是被劈开了一样;如果从公路上看,山体的景色则在一个缺口中若隐若现。因此,建筑本不该在这里出现。或者说,这个项目不仅融入了大地,更融于周围的景观。

Korea DMZ Peace and Life Valley

Architecture of the Land

The unusually named Korea DMZ Peace and Life Valley belongs to a corporation whose slogan is "open the door of peace with the key of life." Noting how the Demilitarized Zone (DMZ), a remnant

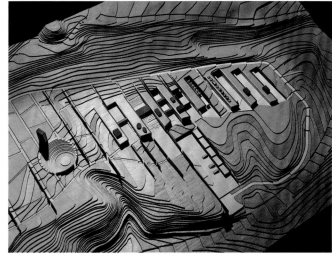

of the war, paradoxically became a scene of life restoration, this organization was founded to campaign for contemplating this era as well as protecting the sanctity of life. The extensive support from Gangwondo and aspiring scholars who formed the key members ensured that this organization was established.

This facility serves as the base camp for DMZ surveys, but it also organizes lectures, seminars, and meditations and includes an agricultural movement to promote life restoration. Thus, the facility itself must maintain a close relationship with nature. From the onset, the project was not allowed to be hostile toward or controlling of nature.

The site is located on a mountain in the town of Inje, which is close to the DMZ. This mountain has a plateau in the middle that faces the west. This area connects the natural mountain range to the carriageway, thereby forming a vital element. Architecture serves as the medium linking all the elements; consequently, it had to be both an artificial nature and a natural artificiality. If viewed from the mountain, it appears as though the land has been cleaved; if viewed from the road, the scenery of the mountain comes in through the gap. Thus, the architecture here should not exist. In other words, the project becomes architecture of the land and of the landscape. Seung, H-Sang

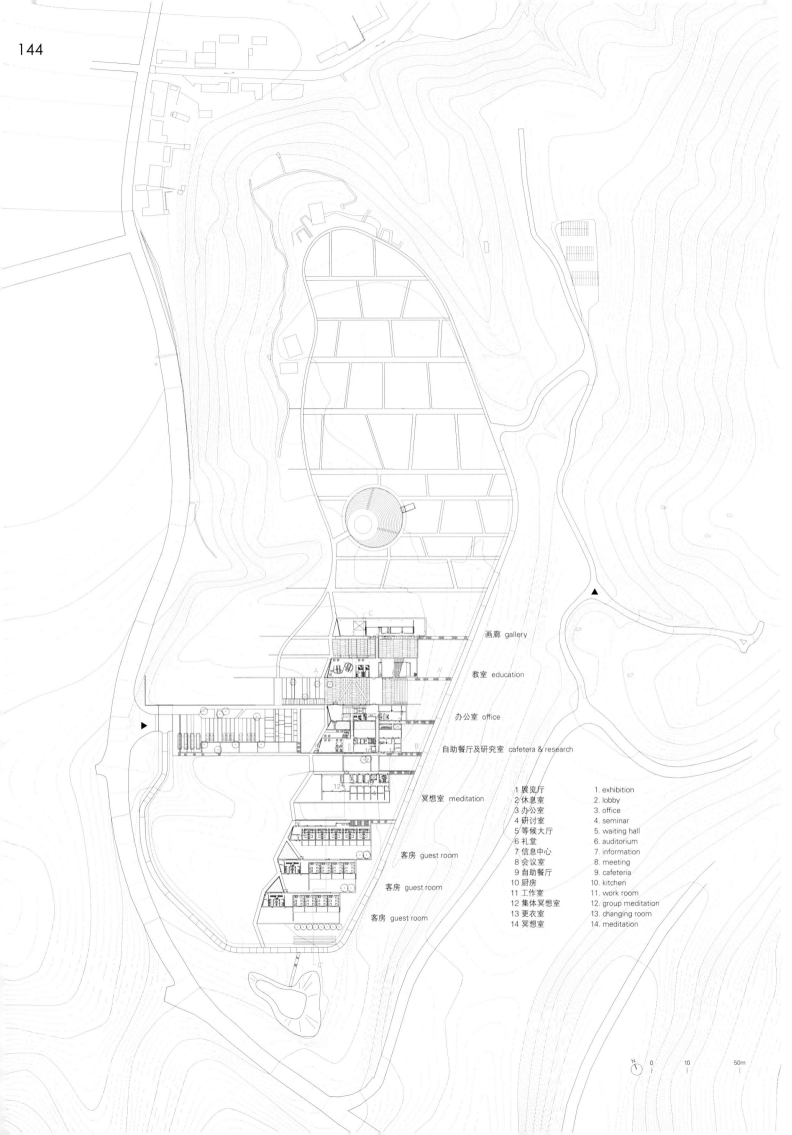

项目名称：Korea DMZ Peace and Life Valley 地点：Inje, Gangwon-do, Korea
建筑师：Seung, H-Sang 项目团队：JongTae Yun
结构工程师：Seoul Structural Eng. 机械工程师：Hanil Mec.
电气工程师：URim Elec. 景观设计师：Landscape Design Studio
项目功能：cultural facilities, educational facilities, accommodation facilities
用地面积：124,210m² 建筑面积：3,275.73m² 总建筑面积：3,304.35m²
建筑规模：one story below ground, one story above ground
结构：reinforced concrete 外部材料：corten steel, soil, pair glass, wood
建造时间：2007.3~2009.8 摄影师：©JongOh Kim

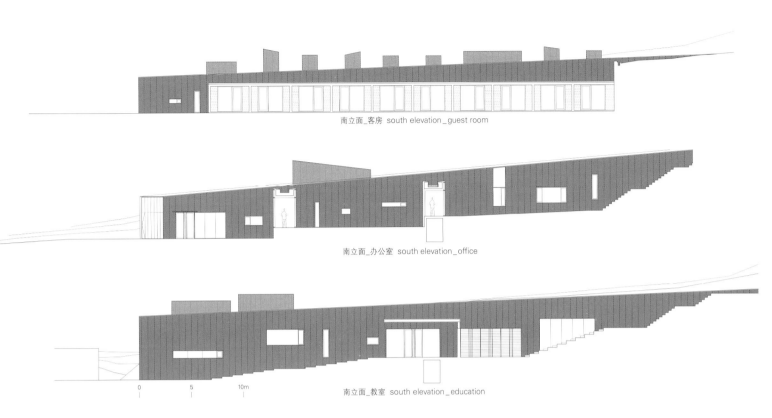

南立面_客房 south elevation_guest room

南立面_办公室 south elevation_office

0 5 10m

南立面_教室 south elevation_education

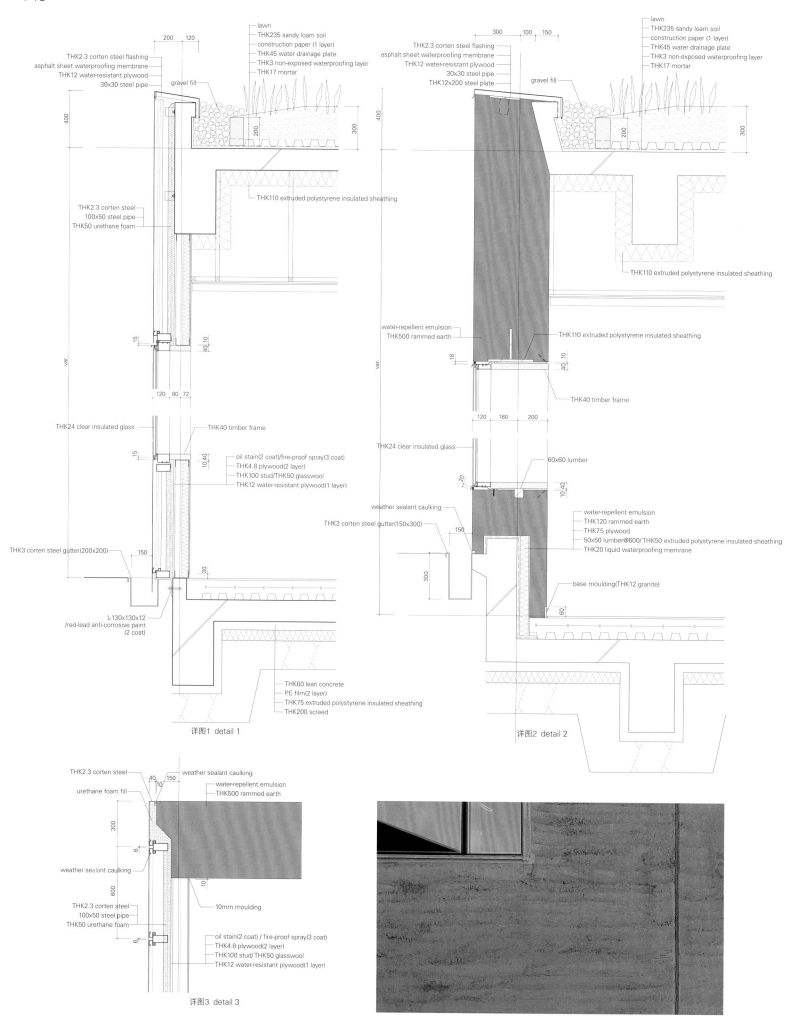

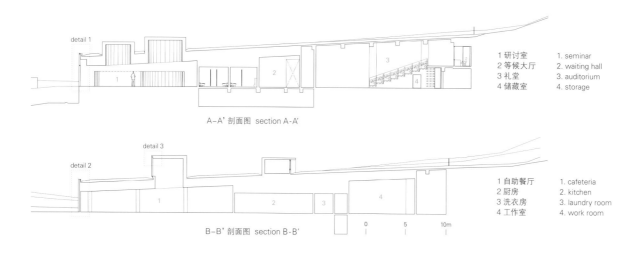

	1 研讨室	1. seminar
	2 等候大厅	2. waiting hall
	3 礼堂	3. auditorium
	4 储藏室	4. storage

A-A' 剖面图 section A-A'

	1 自助餐厅	1. cafeteria
	2 厨房	2. kitchen
	3 洗衣房	3. laundry room
	4 工作室	4. work room

B-B' 剖面图 section B-B'

1 展览厅 2 研讨室 3 休息室 4 商店 5 自助餐厅 6 集体冥想室 7 私人房间

1. exhibition 2. seminar 3. lobby 4. shop 5. cafeteria 6. group meditation 7. private room

C-C' 剖面图 section C-C'

承孝相 Seung, H-Sang

济州岛艺术别墅社区中心

景观中的建筑

将济州岛的景色描绘得最好的地图就是金正浩（朝鲜地理学家）绘制的《大东舆地图》。金正浩在地图上画出了汉拿山的山脊线以及蔓延至整个海洋的山谷线，向人们展示了如何从垂直方向而非水平方向理解济州岛地形。汉拿山和海洋之间的联系是济州岛的生态脊柱，因此，任何破坏这种联系的建筑都是在破坏济州岛。516号公路环绕济州岛一圈，切断了济州岛的生态脊柱。这条滨海大道是一种非生态的公共设施，它使济州岛看起来像一个漂浮在沥青中而非海洋中的小岛。济州岛的基础建筑用通道将汉拿山和海洋连接在一起；这个建筑是沿一条线垂直排列的，使之能够独占看海的视野，这不仅妨碍了自然景物之间的联系，也破坏这个地方的自然环境。

尽管负责这个项目总体规划的DA建筑事务所肯定了本案建筑师的一些观点，但是协商的局限性还是很明显。首要的是对于密度的控制不到位，使我们很难营造度假胜地的轻松环境。因此，对于空走廊的使用变成了关键。

本案建筑师负责的是A地块，该地块位于这一地区的最高处。房子之间的距离被最大化，避免破坏汉拿山的景色。因此，有必要创造一个长而垂直的平面。每个单元都由一个单一的、不分区的房间平面组成，来延续这种流动感。从总体来看，连接汉拿山和海洋的景观在这些单元之间流动，整座房子则保持着一连串的景观视野，而景观通道则成为最引人瞩目的焦点。

俱乐部位于这组综合设施的入口处，保持了风景的主旋律。它坐落在不同的标高上，并在不同标高上伸向不同方向。这些不同的标高通过建筑有机地连接在了一起。在某个区域，地面被用作俱乐部的屋顶平台。沿着倾斜的入口可以到达其他平面；这条路也可以一直通到自助餐厅、院子、画廊以及游泳池。通过这种方式，建筑与地形融为一体。这座建筑既没有突出的外形也没有象征意义，它给人唯一一持久的印象就是景观。

Jeju Art Villas Community Center

Architecture of landscape

The map that best highlights the scenery of Jeju Island is Dae-dong-yuh-ji-do, by Kim Jung Ho. Kim's map draws the mountain ridge of Mt. Hanla and the valley lines spreading out across the ocean, demonstrating how the topography of Jeju Island should be understood as vertical, not horizontal. The connection between Mt. Hanla and the ocean is the spine of Jeju's ecology; thus,

1 餐厅
2 屋顶花园

1. restaurant
2. roof garden

二层 second floor

1 咖啡厅
2 厨房
3 卡拉OK
4 大厅
5 餐厅
6 储藏室
7 便利店
8 办公室
9 高尔夫车存放处
10 员工自助餐厅
11 安全设备室
12 布草房
13 计算机机房

1. cafe
2. kitchen
3. karaoke
4. hall/lobby
5. restaurant
6. storage
7. convenience store
8. office
9. cart parking
10. staff cafeteria
11. safety
12. linen room
13. computer room

一层 first floor

1 室外游泳池
2 储藏室
3 健身房
4 发电室
5 办公室
6 电气室
7 大厅
8 室内高尔夫球场
9 机房
10 信息技术平台

1. outdoor pool
2. storage
3. gym
4. generation room
5. office
6. electricity room
7. hall
8. indoor golf
9. machine room
10. PIT

地下一层 first floor below ground

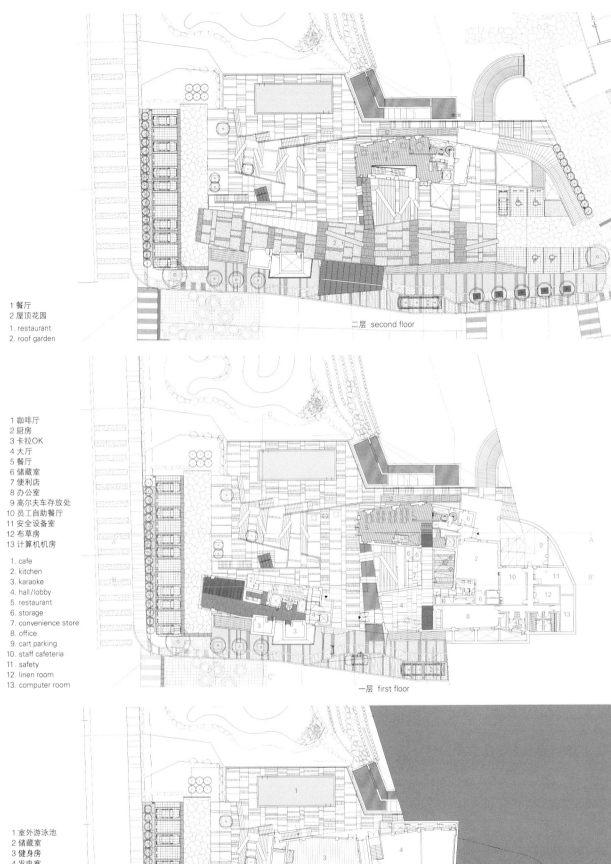
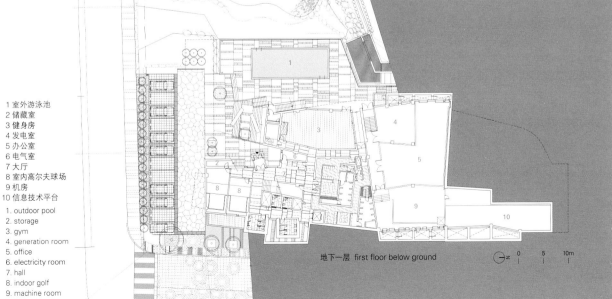

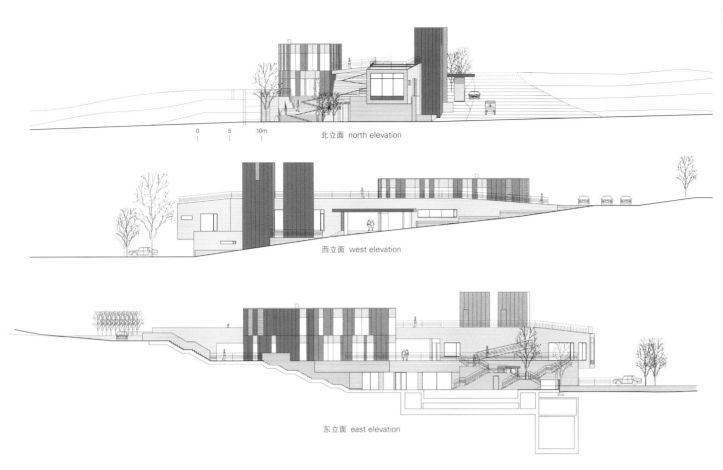

北立面 north elevation

西立面 west elevation

东立面 east elevation

any architecture that interferes with this link would be in opposition to Jeju. Road 516 follows the perimeter of Jeju, severing the spine of the island's ecology. The coastal road, which is a non-ecological public work, makes Jeju seem like an island floating in asphalt, not the ocean. Jeju's fundamental architecture creates a passage connecting Mt. Hanla to the ocean; the architecture is arranged vertically to this line as arranging it to monopolize the view of the ocean not only impedes the flow of nature, but also violates the public nature of the site.

Although DA, which was in charge of the project's master plan, acknowledged quite a few of my opinions, the limitation of the consultation was evident. Above all, the control of the density was not smooth, making it difficult to create the relaxed atmosphere of a resort town. Consequently, the use of empty passages became critical.

I was in charge of Block A, which is located in the highest part of the region. Gaps between households were maximized so as not to interfere with the view of Mt. Hanla. As a result, it was necessary to create a long vertical plane. A unit was made from the flat plane of a single, non-partitioned room to continue the flow. Overall, the scenery connecting Mt. Hanla and the ocean flows between the units as well as the sequence of the scenery is maintained throughout the house, with the passage of the scenery being the main focus.

The clubhouse located at the entrance of the complex maintained the theme of the scenery. The clubhouse sits upon varying levels of ground extending in all directions. These different levels are organically connected via the architecture. In one area, the ground serves as a roof terrace for the clubhouse. Following up the inclined entrance leads to other grounds; the path continues on to the cafeteria, the yard, the gallery, and the pool. In this way, the topography forms the architecture. The building has no form or symbolization. The only lasting impression is that of the landscape. Seung, H-Sang

项目名称：Jeju Art Villas Community Center
地点：SaekDal-Dong Mountain, SeoGwiPo-Si, JeJu Special Self-Governing Province, Korea
建筑师：Seung, H-Sang
项目团队：JongTae Yun
结构工程师：Mirae ISE　机械工程师：Yungdo ENG
电气工程师：JungMyoung　照明设计师：New Lite
景观设计师：HAIIN Land+Scape Design
用途：condominium
用地面积：83,841.60m²　建筑面积：16,173.09m²　总建筑面积：19,211.62m²
建筑规模：62 Villas, 73 Rooms
结构：reinforced concrete
建造时间：2010.7~2012.3
摄影师：©JongOh Kim

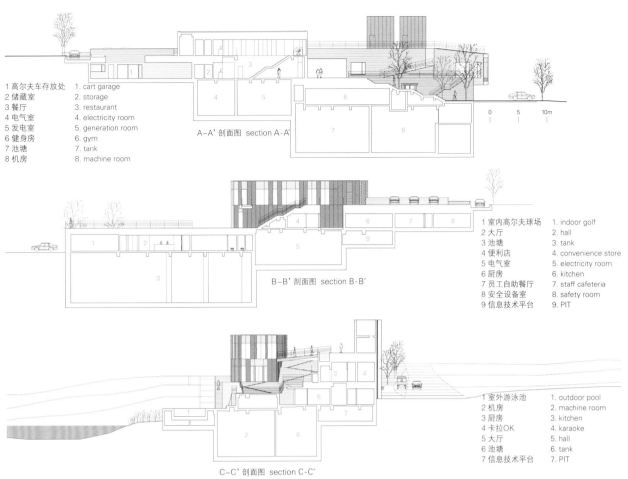

1 高尔夫车存放处	1. cart garage
2 储藏室	2. storage
3 餐厅	3. restaurant
4 电气室	4. electricity room
5 发电室	5. generation room
6 健身房	6. gym
7 池塘	7. tank
8 机房	8. machine room

A-A' 剖面图 section A-A'

1 室内高尔夫球场	1. indoor golf
2 大厅	2. hall
3 池塘	3. tank
4 便利店	4. convenience store
5 电气室	5. electricity room
6 厨房	6. kitchen
7 员工自助餐厅	7. staff cafeteria
8 安全设备室	8. safety room
9 信息技术平台	9. PIT

B-B' 剖面图 section B-B'

1 室外游泳池	1. outdoor pool
2 机房	2. machine room
3 厨房	3. kitchen
4 卡拉OK	4. karaoke
5 大厅	5. hall
6 池塘	6. tank
7 信息技术平台	7. PIT

C-C' 剖面图 section C-C'

"MoHeon" 与 "SaYaWon"

"MoHeon" 与 "SaYaWon"

在整个建造工程临近结束的时候,业主Yoo Jaesung先生让我来给这座小房子命名。他自己已经是一家钢铁制造厂的董事长了。此时,有一个词进入了我的脑海,那就是"MoHeon",意思是"某人的房子"。"MoHeon"的含义是一座无名的、不存在的房子。在Yoo先生欣然接受了这个名字以后,一块由中国书法家写的"MoHeon"的牌子就挂在了房子里。这座房子本就该以此为名。

新建的"MoHeon"与Yoo先生35年前建成的房子相连。在同一个城市的同一座房子里居住35年绝不是一件容易的事。要说某人在这一段时间都住在同一个地方是不大可能的,特别是在韩国。因为韩国在这35年间环境、生活方式和人们的心态都发生了很大的改变。刚开始的时候,庆北大学的前门就在这个地方。随着大学的次入口替代了前门,这座房子周围的景观就被破坏了。但是,Yoo先生想要通过建筑设计和景观设计来恢复被破坏的地区。他的首要任务是要买下场地对面的部分,然后通过建造传统的日式花园来使它变得独特。在和一位传统的日本景观设计师合作设计这一项目的同时,Yoo先生还买下了与这块场地相邻的一块地,想要建造一个韩式的花园。我受聘成了这个韩式花园的设计师。

我认识Yoo先生已经有10年了,但是我们还是需要相互了解。鉴于他了解并接受我鲜明的个性,所以除了一些基本的功能要求外,他把所有的事情都交给我来决定。当我推荐Jeong YoungSun女士来做这个韩式花园的景观设计的时候,Yoo先生看起来不太情愿接受,因为对于景观设计他有自己的想法。但是,他极力尊重我的决定。

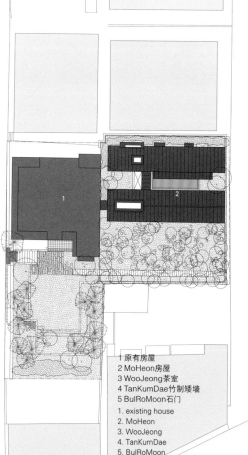

1 原有房屋
2 MoHeon房屋
3 WooJeong茶室
4 TanKumDae竹制矮墙
5 BulRoMoon石门
1. existing house
2. MoHeon
3. WooJeong
4. TanKumDae
5. BulRoMoon

我对这个项目的最初设计理念是把客房作为附属空间，与主体建筑相离。如果这样，这个项目的主体就是一间客房和一个能够为客人提供聚会和餐饮场所的、充满欢乐的花园。如果附属空间的功能是客房，那么我认为它需要隐匿在自然环境里。我们的首要任务是尽量使建筑的空间丰富和多样化，并且应该对周围空间起到限定的作用。最终，我设计了四个庭院。为了尽量放大前庭空间，建筑只得向后退，在空间上形成两个体块。第一个体块是一个餐厅，我将它设计成了透明的，它的通透性使得视线可以从中穿过。因此，从卧室所在的第二个功能体块就可以看到前庭。在这两个体块之间，设置了两个不同形式的庭院。一个是水园；一个是下沉的庭院，使得自然光线能够进入地下。从卧室的窗户向外还可以看到一个小竹园。这样的视觉联系使空间显得更加深远和丰富。餐厅里带窗户的墙体是可以移动的，使我们可以创造出更加多样化的空间。对于这个通透的餐厅来说，重要的是能够通过它有一个视觉上的延伸。整个场地都被与房子高度相同的耐候特种钢墙体包围着，进一步加强了空间的延伸感。设计好了这座房子后，我开始等待Jeong Youngsun的景观设计。

Jeong女士的景观设计远远超出了我和Yoo先生的预料。我本打算在设计中放大庭院的尺寸，但是它只有165.3m²。因此，我本来对花园的设想是，在耐候特种钢墙体的背景下，庭院里屹立着粗壮的白色树干。但是Jeong女士告诉我们，她打算在这样一个小院子里密集地种植树木并在地面上铺满小碎石。我仍然记得我去场地看景观时的那一刻它给我带来的强烈的冲击。一条笔直的石子路连接着主体建筑和庭院。我一踏上那条石子路，就有一种很强烈的融入到了森林里的感觉。真是太神奇了！我的目的达到了，建筑的确隐没在环境里，而只有景观是我们所留意、所感受到的东西。如此壮美的景象在一个这样小的空间里得到了彰显，的确很令人惊讶。而且，它与那个日式花园大不相同。

在安装着篷式天窗的客房中充满了安逸和宁静的感觉。穿过水园和透明的餐饮空间，由耐候特种钢墙体围起来的铺有石子的花园映入眼帘。不仅如此，还能听到雨水淅沥沥落下和微风飒飒作响的声音、下雪时雪花漫天飞舞然后将大地覆盖的声音……我希望Yoo先生也会彻底喜欢上这个空间。

另一方面，日式花园在设计和建造上都使用了传统的日本方式，现在正处在最后的施工阶段。这个日式花园的地面上长满了苔藓。

最初的设想是建造一间日式的茶室，但是Yoo先生在看到"MoHeon"展示出来的效果以后，改变了设计与施工的方向。他说服了那位日本景观设计师，让我来设计这个茶室。这个决定是正确的，因为如果不是这样的话，这个日式花园就会成为特别突兀而又孤立的景观。在这里就应该建造一座具有"MoHeon"含义的当代建筑。这样，这块场地就会作为统一的整体，即一部完整的作品来影响我们这个时代。

日式花园里的建筑也是附属部分。茶室要随着时间的推移而有所变化，甚至有时候将自己藏匿起来。所以，门窗是三层的。窗户由韩纸（一种传统的韩国手工纸）、玻璃和木质百叶窗三层构成。由这三层构成的茶室的形态在时时变化着，有时候甚至会藏匿起来，仿佛消失了一般。Yoo先生请我为这个茶室命名，我给它取名为"WooJeong"，意思是"不存在"。

我设计了竹制的矮墙以取得和另一边池塘的视觉上的连通，这面

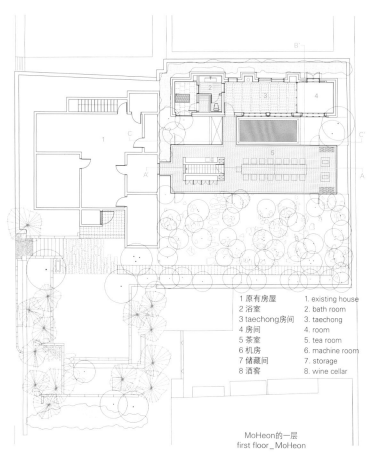
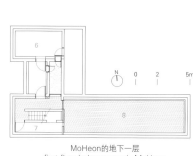

1 原有房屋	1. existing house
2 浴室	2. bath room
3 taechong房间	3. taechong
4 房间	4. room
5 茶室	5. tea room
6 机房	6. machine room
7 储藏间	7. storage
8 酒窖	8. wine cellar

MoHeon的一层
first floor _ MoHeon

MoHeon的地下一层
first floor below ground _ MoHeon

项目名称：MoHeon and WooJeong
地点：SanGyuk-dong, Buk-gu, DaeGu, Korea
建筑师：Seung, H-Sang
项目团队：DongHee Lee
结构工程师：eoul Structural Eng.
机械工程师：Seah Eng.
电气工程师：URim Elec.
照明设计师：Bitzro
景观设计师：Seoahn Total Landscape
用途：Tea House
建造时间：2009.11—2011.11
摄影师：©JongOh Kim

MoHeon
用地面积：357.16m² 建筑面积：120.90m²
总建筑面积：164.68m² 高度：7.45m
建筑规模：one story below ground, one story above ground
结构：reinforced concrete
项目用途：Housing
外部材料：THK24 pair glass, THK2.3 corten, wood,
内部材料：exposed mass concrete, wood floor

WooJeong
用地面积：1,050.02m²
建筑面积：15.94m²
总建筑面积：15.94m²
建筑规模：one story
结构：steel frame

北立面 north elevation 0 2 5m

A-A' 剖面图 section A-A'

B-B' 剖面图 section B-B'

C-C' 剖面图 section C-C'

竹制的矮墙叫作"TanKumDae"。在"TanKumDae"上没有活动的时候,"WooJeong"和"TanKumDae"仍然在某种程度上相互联系。这个花园在视觉联系方面设计得非常好。

　　实际上,重要的并不是将所有的元素都安置在恰当的地方,而是通过不同地方、不同元素之间的相互关系使得它们都变得清晰易读。于是,在这里的每一个地方,看到这样的景色,可以产生很多种不同的理解方式。有着粗壮树干的松树和旧式的石匠建造的石塔能很好地融入这个日式花园的氛围。

　　出于尊敬,应该为这样一座与众不同的花园设计一个小门。建造前总统卢武铉纪念碑的两位主要石匠Yoon和TaeJoong负责设计这个小门。这个小门是由石头建造的,这块石头Yoo先生收藏了很长时间。小门的名字叫作"Bulromoon"。

　　当整体施工快结束的时候,Yoo先生委托Min KyungSik对现有的住宅做修复。他对原有的墙体进行了自由调整并且将这座老房子融入了一个可以同时欣赏到日式花园和"MoHeon"的美妙环境中。

　　最终,这三部分区域——日式园林、原先的主体建筑以及"MoHeon"——都和谐一致地融为一体。

　　Yoo先生把整个项目命名为"SaYaWon"。我并没有问他SaYa这个词的含义,但是我猜测,它也许是意味着"自我放逐的学者"。

　　"这是一群自愿与社会脱离关系、不愿循规蹈矩但又无惧艰难险阻的人,也是一群勇于改变的人。"爱德华·萨义德曾经说过这些人都是智者。如果这样的话,那么Yoo先生无疑正是一位智者。

　　Yoo先生长期以来一直在资助一批有天赋的年轻艺术家。他在环游世界时接受了各国的文化熏陶,同时他也对贫瘠的现代韩国文化和艺术拥有深厚的感情。但是他经常站在幕后,处事低调,并非公众人物。因此,他的房子就应该被命名为"MoHeon"。

MoHeon and SaYaWon

When the construction was approaching completion, the client, Yoo Jaesung, who became the chairman of a steel manufacturing company by himself, asked me to name this tiny house. At this moment, a word popped into my head and it was "MoHeon", meaning someone's house. "MoHeon" means a non-existent house with no name. Right after Mr. Yoo willingly consented to this naming, the letters of "MoHeon" were placed in this house written by a Chinese calligrapher. This house must be named thus. "MoHeon" is attached to Mr. Yoo's house, which was built 35 years ago. It is not easy at all to live in a house in a city for 35 years. It is not believable that for this period of time he has lived in the same place, especially in Korea where circumstances, lifestyle, and psychology have been changing inexorably for three and a half decades. Originally, the front entrance of KyungBook University was in this area. As the sub-entrance replaced the front, the surrounding landscape of this house was destroyed. However, Mr. Yoo intended to recover this destroyed area through architecture and landscaping. His first task was to buy the opposite side of the estate and start to make it special by creating a traditional Japanese garden. While working on the project with a traditional Japanese landscape architect, Mr. Yoo also bought an estate right next

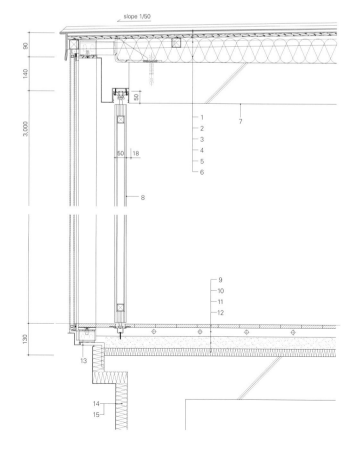

1. THK100 extruded polystyrene insulated sheating
2. □-45x45 ST'L pipe@1,500
3. THK12 water-resistant plywood (1 layer)
4. THK1 reflective thermal insulation sheet
5. THK10 ventilation mat
6. THK0.7 titanium double-standing seam
7. exposed concrete
8. THK6 birch plywood folding door
9. THK14 wood flooring for on-dol
10. THK36 cement mortar (w/xl-pipe)
11. THK50 ALC
12. THK30 thermal insulation
13. THK10 150x90/anti-corrosive paint/PVDF coating
14. THK1.2 150x90/anti-corrosive paint/PVDF coating
15. THK50 extruded polystyrene insulated sheating
16. THK9 water-resistant plywood (2 layer)
17. TKK1.2 oil paint in galvanized steel
18. THK1.6 galvanised flashing/anti-corrosive paint /PVDF coating
19. THK2 AL. sheet
20. door hanger
21. THK24 insulated glass
22. THK1.6 glavanised steel/oil paint
23. □-150x100 OSB
24. putty/acrylic paint
25. THK9.5 gypsum board
26. □-30x30 lumber
27. clear lacquer spray on red pine
28. korean traditional style Hna-Ji top-hung window
29. □-300x200 laminated timber
30. urea foam fill

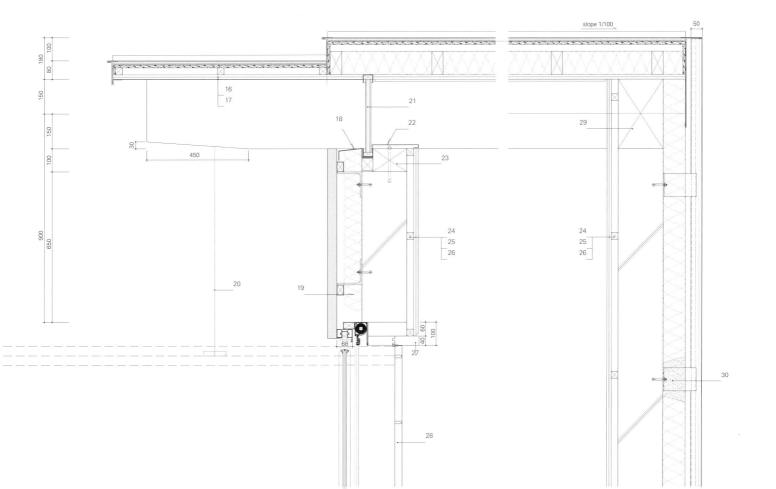

to the existing house in order to create a Korean garden and I was commissioned to design this project.

It had been 10 years since I met Mr. Yoo, but we still needed to get to know each other. Since he knew and accepted my sharp personality, he left everything to me saving some basic required functions. When I recommended that Ms. Jeong YoungSun designed the landscaping for the Korean garden, Mr. Yoo seemed to consent reluctantly, because he had his own ideas about the landscaping. However, he tried his best to respect my decisions.

My initial proposal for this project was to design the guest room as an attached space off of the main house. Therefore, the main programs of the house are a guest room and a space where people can have a party or dinner in a joyful garden. If the aim of the attached space was to be a guest room, I thought its existence should be veiled and concealed. The priority was to make the spaces of this house as various and rich as possible, and the house itself should be a tool for defining spaces. Eventually, four courtyards are created. To maximize the space for the front courtyard, the building had to be set back and organized as two spatial layers. I made the first layer a translucent dining room, thereby allowing a view of the front courtyard through it from the bedroom (the second layer). Between these two layers, two kinds of courtyards are created. One is a garden of water and the other is sunken to allow light underground. From the bedroom, a small bamboo garden can be seen over the window. This kind of visual connectivity makes the space much deeper and richer. The dining room has movable walls that are attached to windows and allow one to create various spaces. The important issue with the dining room was to have visual extensions through its transparency. The entire site is surrounded by corten-steel walls at the same height as the house's and the encloser emphasizes the tension of the space. After designing the house, I waited for Jeong Youngsun's landscaping.

Ms. Jeong's landscaping was much more than Mr. Yoo and I had expected. I was planning to maximize the garden area, but it was only 165.3m². Therefore, what I was hoping to see were a few white and thick tree trunks soaring, the stream of corten-steel walls behind. Ms. Jeong informed us that she planted loads of trees densely in such a tiny space and filled the ground with small rubble. I still remember the moment and its strong impression when I visited the site to see the landscaping. There is a straight path made of stone pavers connecting the main house to this courtyard. As soon as I stepped on the stone paver, I felt as if I had been sucked into the massive woods. It was magical. My expectation that the architecture should be concealed and only the landscape should be noticeable was realized. That such a sense of beauty

and the sublime could be manifested in such a small space was so surprising, and it was so different from the Japanese garden.
Being in the guest room, filled with its profound peace and quiet, awning windows, walking through the water garden and transparent dining space, the scenery of the stone garden enclosed by the stream of corten-steel walls and further, the sound of the rain and breeze, the sound of snow covering the ground… I hoped Mr. Yoo too would love this space with all his senses. The Japanese garden on the other hand, designed and constructed in the traditional Japanese method, was now under its last phase of construction, being covered with moss.

The initial idea was to build a Japanese tea house but Mr. Yoo changed the direction of the construction and design upon seeing "MoHeon" manifest. He convinced the Japanese landscape architect that I should design the tea house instead. It had to be done this way, otherwise, the Japanese garden would have remained as a chauvinistic isolated landscape. Contemporary architecture relating to "MoHeon" had to be in this site. In this way, our current era would be reflected by the entire site as a merged, unified, completed work.

The architecture in the Japanese garden is also an additional part. The tea house is intended to be changed from time to time, and even to be veiled sometimes. That is why windows and doors are comprised of three layers – windows of Hanji(traditional Korean handmade paper), glass and wooden louvers. By the composition of these three layers, the tea house has different appearances every single time and sometimes is concealed and vanishes. Mr. Yoo also asked me to name this tea house and I called it "WooJeong", or non-existence.

The bamboo podium is designed to have visual connections to the pond on the other side and this podium is called "Tankumdae". When there are no events on "TanKumDae", "WooJeong" and "TanKumDae" still communicate with each other somehow. This garden is well-designed in terms of the visual connectivity.

It is not important that objects themselves are placed in appropriate locations, but the important issue is that the relationship between objects in different places allows each of their specific stories to be read legibly. Thus, everywhere in this place, diverse stories can be understood by looking at the landscape. The pine tree with the massive trunk and the pagoda built by a stonemason in the old days are well-harmonized with the atmosphere of the Japanese garden.

It was necessary to have a small gate out of respect for such an extraordinary garden like this. Yoon, TaeJoong, who was the main stonemason for the memorial for the former president Roh, Moohyun, was in charge of the gate, and this gate is made out of stone, which Mr. Yoo has been collecting for a long time. It is named "Bulromoon".

When everything reached completion, Mr. Yoo commissioned Min, KyungSik to repair the existing house. He adjusted the old walls freely and made the existing house into a space for enjoying and viewing both the Japanese garden and "MoHeon".

Eventually, those three areas – the Japanese garden, the main house, and "MoHeon" – were harmonized and merged together and three parts were combined into one.

Mr. Yoo named the entire project "SaYaWon". SaYa… I did not ask him about the meaning of it, but I guess, it might mean "voluntarily-exiled scholar."

"a person who is voluntarily renouncing society, a person who is not responding with conventional logic but responding with boldness and adventurous courage, a person who is representing and responding to changes." Edward Wadie Said said that such people were intellectuals. If so, Mr. Yoo is definitely an intellectual.

Mr. Yoo has supported young talented artists for a long time. While he travelled around the world, he was culturally enriched, but at the same time he has a deep affection for the barren culture and art of modern Korea. And he is always standing behind the scenes and out of the spotlight. Therefore, his house should be named "MoHeon". Seung, H-Sang

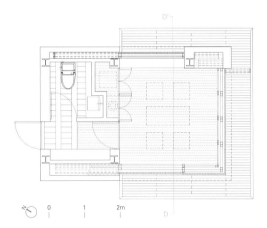

WooJeong的一层 first floor_WooJeong

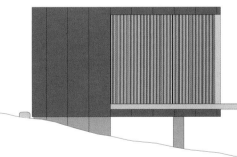

WooJeong的东立面 east elevation_WooJeong

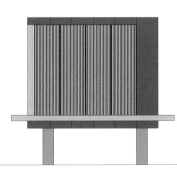

WooJeong的北立面 north elevation_WooJeong

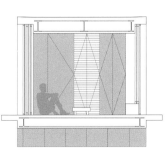

D-D'剖面图 section D-D'

>>120
Seung, H-Sang
Graduated with his master's degree from Seoul National University and studied at Vienna University of Technology. After working for SwooGeun Kim from 1974 to 1989, established Iroje Architects & Planners in 1989. In 2002, was the first architect to be named Artist of the Year by the National Museum of Contemporary Art, Korea, where he held a grand private exihibition. Was a visiting professor of North London University and taught at Seoul National University and at Korea National University of Art. In 2007, Korean government honored him with "Korea Award for Art and Culture", and was commissioned as the director for Gwangju Design Biennale 2011 after for Korean Pavilion of Venice Biennale 2008. Currently has been invited to Venice Biennale 2012.

>>58
Naf Architect & Design
Tetsuya Nakazono was born in Miyazaki, Japan in 1972. Graduated from Hiroshima University, department of architecture in 1995. Received a master's degree in 1997 and worked for Shiomi Architects and Associates. Established Naf Architect & Design in 2001. Currently works as a representative director of Naf Architect & Design and assistant professor in Sojo University.

>>46
Takeshi Hosaka Architects
Takeshi Hosaka was born in Japan, in 1975. Received a degree in architecture in 1999 and a master's degree in 2002 at Yokohama National University. Established Takeshi Hosaka Architects in 2004. Currently is lecturing at Kokushikan University, Hosei University, and Yokohama University.

>>32
Suga Atelier
Shotaro Suga was born in Osaka, Japan in 1956. Graduated from Kyoto Institute of Technology in 1980 and established Suga Atelier in 2001. Currently teaches at several universities including Osaka City University and Kobe Design University.

>>22
mago
is an architectural design and engineering firm located in Braga, Portugal.
António Jorge Fontes[left] received a degree in civil engineering at the University of Minho in 1992 and in architecture at the School of Arts in Porto in 2000. Completed a master's degree in municipal engineering, with an option in municipal town planning in 2005. Began teaching at the University of Minho in 2002, and also taught construction at the Artistic High School of Porto. In 2009, started PhD in construction process.
André Fontes[right] received a degree in civil engineering at the University of Minho in 1992 and also in architecture at the School of Arts in Porto in 1999. Granted a master's degree in municipal engineering, with a dissertation entitled "Urban Environment for Pedestrians" at the University of Minho in

>>68
Alberto Campo Baeza
Was born in 1946, in Spain. Graduated from Madrid Technical School of Architecture(ETSAM) in 1971. Has been teaching at several universities and institutes. Has ever completed a number of significant projects. His works have been published in major architectural magazines and exhibited in major cities. Most recent works are Cultural Center in Cobquecura, Chile and Olnick Spanu Museum in New York.

BongHee Jeon
Studied at Seoul National University and received a doctor's degree with the study on Korean architecture history. Currently works as professor at the department of architecture, Seoul National University. As well as establishing archives on traditional Korean housing and its modernization, has performed field surveys and projects in East Asian countries including China, Japan and Vietnam. His subjects vary from ancient Korean architecture to modern and contemporary architecture. Has published *An Architecture Survey of Beijing Siheyuan in Xijiulianzi Hutong; 3×3, a typological Approach on Korean Architecture* and a translation work, *Holzbaukunst*.

Diego Terna
Is an architect and curator of the section "Imagined architectures" in the webzine PresS/Tletter. After winning a young critics competition, called by Italian critic Luigi Prestinenza Puglisi, his writings were published in several art and architectural magazines (Compasses, Arch'it, Exibart). Has been working for the office of Italo Rota since 2005, after an experience at Boeri Studio.

Michele Stramezzi
Received a master's degree in architecture and urbanism at Polytechnic of Milan School of Architecture in 2003. Has ever worked at a number of offices in the Netherlands, amongst others, de Architekten Cie, Erick van Egeraat associated architects and MVRDV as a project leader. After moved to Beijing, China, has been working as a freelance architect and architectural consultant.

Maria Pedal
Is a German architect and urbanist. Received her master's degree in urbanism at Technical University Delft(TU Delft). Worked in several Dutch offices in Amsterdam and Rotterdam. Has been living and working as an urban planner and designer in Beijing, China since 2009.

>>98
Carmen Izquierdo
Was born in 1974 in Madrid, Spain. Graduated from Madrid Technical School of Architecture(ETSAM) in 2000. Worked for Mariano Bayón Arquitectura y Urbanismo as a project architect from 2000 to 2003 and established her own office in 2002.

>>104
Gonzalo Moure Architect
Gonzalo Moure Lorenzo has been working as an architect since 1986. Has won a lot of architectural competitions of national and international scope. Most of his works, built or currently under construction, are the results of architectural competitions. Has been a professor at the architectural design department of Madrid Technical School of Architecture(ETSAM) since 1991. Many of his works have been published in national and international journals.

>>86
Hackett Hall McKnight
Is an architectural practice based in Belfast. The partnership of Alastair Hall[right] and Ian McKnight[left] countinues following the retirement of Mark Hackett from the practice in June 2010. Alastair Hall studied architecture at Queens University in Belfast and Cambridge University. Upon graduation worked for Grafton Architects in Dublin before returning to Belfast. Has taught in Queens University, Belfast and has been a guest critic at several universities in Ireland. Ian McKnight studied architecture at the University of Newcastle and the Mackintosh School in Glasgow. His early career was spent in London where he worked for David Chipperfield Architects. After his return to Belfast in 2000, worked as an associate in a local practice, winning several design contests and completing a number of significant projects and winning a number of national awards.

图书在版编目(CIP)数据

内在丰富性建筑：汉英对照 / 韩国C3出版公社编；王思锐译. —大连：大连理工大学出版社，2012.11
（C3建筑立场系列丛书；21）
ISBN 978-7-5611-7444-9

Ⅰ.①内… Ⅱ.①韩…②王… Ⅲ.①室内装饰－建筑设计－汉、英 Ⅳ.①TU238

中国版本图书馆CIP数据核字（2012）第269739号

出版发行：大连理工大学出版社
　　　　　（地址：大连市软件园路80号　邮编：116023）
印　　　刷：精一印刷（深圳）有限公司
幅面尺寸：225mm×300mm
印　　张：11.5
出版时间：2012年11月第1版
印刷时间：2012年11月第1次印刷
出 版 人：金英伟
统　　筹：房　磊
责任编辑：张昕焱
封面设计：王志峰
责任校对：张媛媛

书　　号：ISBN 978-7-5611-7444-9
定　　价：228.00元

发　行：0411-84708842
传　真：0411-84701466
E-mail: 12282980@qq.com
URL: http://www.dutp.cn